Ecofeminism and the Indian Novel

Ecofeminism and the Indian Novel tests the theories of ecofeminism against the background of India's often different perceptions of environmental problems, challenging the hegemony of Western culture in thinking about human problems. This book moves beyond a simple application of the concepts of ecofeminism, instead explaining the uniqueness of Indian novels as narratives of ecofeminism and how they can contribute to the development of the theory of ecofeminism.

In examining a selection of novels, the author argues that Indian texts conceptualise the ecological crisis more as a human problem than as a gender problem. The book proposes that we should think of ecofeminism as ecohumanism instead, seeing human beings and nature as a part of a complex web. Novels analysed within the text include Kamala Markandaya's *Nectar in a Sieve* (2009), Shivram Karanth's *Return to Earth* (2002) and Na D'Souza's *Dweepa* (2013).

Ecofeminism and the Indian Novel will be of great interest to students and scholars of ecofeminism, ecocriticism, ecological feminism, environmental humanities, gender studies, ecological humanities, feminist studies and Indian literature.

Dr Sangita Patil is an Assistant Professor at LBS Govt First Grade College, Bengaluru, India. Her research interests include ecofeminism, literary theory, cultural studies and liberal education.

Routledge Explorations in Environmental Studies

Stranded Assets and the Environment
Risk, resilience and opportunity
Edited by Ben Caldecott

Society, Environment and Human Security in the Arctic Barents Region
Edited by Kamrul Hossain and Dorothée Cambou

Environmental Performance Auditing in the Public Sector
Enabling Sustainable Development
Awadhesh Prasad

Poetics of the Earth
Natural History and Human History
Augustin Berque

Environmental Humanities and the Uncanny
Ecoculture, Literature and Religion
Rod Giblett

Ethical Responses to Nature's Call
Reticent Imperatives
James Magrini

Environmental Education in Indonesia
Creating Responsible Citizens in the Global South?
Lyn Parker and Kelsie Prabawa-Sear

Ecofeminism and the Indian Novel
Sangita Patil

For more information about this series, please visit: www.routledge.com/Routledge-Explorations-in-Environmental-Studies/book-series/REES

Ecofeminism and the Indian Novel

Sangita Patil

Routledge
Taylor & Francis Group
LONDON AND NEW YORK

earthscan
from Routledge

First published 2020
by Routledge
2 Park Square, Milton Park, Abingdon, Oxon OX14 4RN

and by Routledge
52 Vanderbilt Avenue, New York, NY 10017

Routledge is an imprint of the Taylor & Francis Group, an informa business

First issued in paperback 2021

British Library Cataloguing-in-Publication Data
A catalogue record for this book is available from the British Library

Library of Congress Cataloging-in-Publication Data
A catalog record for this book has been requested

ISBN: 978-0-367-19833-6 (hbk)
ISBN: 978-1-03-209118-1 (pbk)
ISBN: 978-0-429-24354-7 (ebk)

Typeset in Bembo
by Apex CoVantage, LLC

Dedicated to my beloved and too dear husband, Shiva

Contents

Acknowledgements ix

Introduction: ecofeminism and the Indian novel 1

Introduction 1

Ecofeminism: environmental studies and feminism 1

Ecofeminism: Indian polemical discourse 3

Indian novel 4

1 **An interface between human beings and nature** 11

1.1 Introduction 11

1.2 The genealogy of ecofeminism 11

1.3 Indian polemical discourse: environment and ecofeminism 18

1.4 The genealogy of the Indian novel 23

1.5 The discursive formation of the environment in Indian novel 26

1.6 Rumination on Indian environmental movements and protests 29

1.7 Conclusion: the formulation of framework 32

2 **Narratives of agriculture: *Nectar in a Sieve, The Upheaval, Return to Earth* and *Gift in Green*** 37

2.1 Introduction 37

2.2 Nectar in a Sieve*: the impact of tannery on pastoral life 37*

2.3 The Upheaval*: the impact of mining on farming community 49*

2.4 Shivram Karanth's Return to Earth*: the impact of modernisation on agrarian culture 62*

2.5 Sarah Joseph's Gift in Green*: a toxic discourse 70*

2.6 Conclusion 86

3 **Dam construction and ecological crisis: *The Coffer Dams* and *Dweepa*** 89

3.1 Introduction 89

3.2 The project of dam construction and ecological crisis in India 89

3.3 Kamala Markandaya's The Coffer Dams*: modern juggernaut 91*
3.4 Na D'Souza's Dweepa*: an island of destruction 103*
3.5 Conclusion 110

4 The industrial disaster: *Animal's People* 112
4.1 Introduction 112
4.2 The Bhopal gas tragedy: a backdrop 113
4.3 Patriarchal developmental attitude: industrial disaster 113
4.4 Women as victims of the industrial disaster 120
4.5 Rhetorical tropes 122
4.6 The uniqueness of Animal's People *as an ecohumanist narrative 123*
4.7 Conclusion 125

5 Animals as absent referents: *The Man from Chinnamasta* 126
5.1 Introduction 126
5.2 The mythological background 127
5.3 The ethnography of animal sacrifice 128
5.4 Patriarchy and animal sacrifice 129
5.5 Women's concern for animals 132
5.6 The uniqueness of The Man from Chinnamasta *as an ecohumanist narrative 134*
5.7 Conclusion 136

6 Reconceptualising ecofeminism: from ecofeminism to ecohumanism 138
6.1 Introduction 138
6.2 An overview of the ecofeminists' concern 138
6.3 Reconceptualising ecofeminism: from feminism to humanism 143
6.4 Conceptual framework 145

Index 149

Acknowledgements

My special and indelible thanks to my teacher and mentor Professor NS Gundur, who held my hand as I made my baby steps in the journey of scholarly reading and writing; it is not exaggeration to say that he literally taught me how to read and write by selecting a few passages from Aristotle's *Politics.* He helped me realise true meaning of reading and writing. "Teaching scholarship and leading the life of the mind" (my teacher's ideology, which is quite the enthymematic mantra) is the focal point of my research writing. This book is the legacy of his constant guidance and suggestions of conceptual and logical development of thoughts. In due course, I feel without him this book could not have been borne. These words are very scanty to express my gratitude to his remarkable contribution.

Further, I am deeply indebted to my other teachers Dr Panduranga Rao V and Professor Vijaykumar Suryan for spending several hours on a draft copy of the manuscript in editing and proofreading it. Mr. Sanjay Patil, my brother, found time in his busy schedule to proofread some parts of the manuscript. I am grateful to his constant guidance and encouragement. I am also thankful to my friend Mr. Vinod for his intellectual inputs. I extend my sincere thanks to Ms. Manisha Rao, SNDT Women University, Mumbai, for sending the details of new publications in this domain. I must extend my thanks to Mr. Pavangangadhar, Professor Suverna Patil, Dr Vandana and Mr. Vasant for their support in this endeavour.

I take this opportunity to acknowledge the valuable guidance of Professor Narkhide, Department of Environmental Science; M U Mahavidyalaya, Udgir; Dr Girija Jayashankar, ASC Degree College; Dr Yogananda Rao, Jain University and Dr M S Chaitra, Aarohi Research Foundation, Bengaluru, for going through the manuscript and for providing fruitful suggestions to improve upon it.

Special thanks to Senior Editor Claudia Alvares, Editorial Assistant Paul Laurence and two anonymous reviewers of my article that has appeared in Taylor Francis' (Routledge) Cogent Social Science for their critical inputs and suggestions. The exchange of ideas with them has made me rethink my original conceptual understanding of ecofeminism.

I have immensely borrowed resources from several libraries and educational institutions. I am very grateful to: NMKRV College, Bengaluru; M U Mahavidyalaya, Udgir; The Centre for Women's Studies, University of North Bengal, Darjeeling; The Central Library, University of North Bengal, Darjeeling; Dr. VKRV Rao Library, Institute for Social and Economic Change, Bengaluru; The Central Library, Banaras Hindu University, Varanasi; English and Foreign Languages University, Hyderabad among others.

Finally, I would like to express deep appreciation for my family, especially, my husband, Shivkumar, who has been a constant source of inspiration and has always been with me in all my endeavours. Now, he is no more but this book is his dream project. I dedicate this book to my beloved hubby. . . . I owe all my educational achievements to my parents, especially my mother Laxmibai. Moreover, I cannot forget the love and affection of Shashank, Sanket and Rashmi who made me free from taking care of the household chores. Thank you, the Patil fraternity.

Introduction

Ecofeminism and the Indian novel

Introduction

This book explores the ecofeminist theory in Indian context by analysing select Indian novels; the work follows a similar pattern to Foucault's exploration of the evolution of the conceptual meaning of truth through the review of the fictional narratives of Homer and Sophocles (Gundur 2017). Although it is an ecofeminist reading of the Indian novels, the study is not a simple exercise in applying the concepts of ecofeminist theory to the Indian novels. It is an attempt to engage with the theory of ecofeminism in the light of reading of the Indian novels. Application of theories, following Sheldon Pollock's observations, does not advance knowledge. Instead, we need to test the theories (Pollock 2006). In this line, David Lodge's remark is very apt, "What is essential, however, is that the new theoretical self-consciousness should be earned, not borrowed, that it should be based on a study of the seminal texts that gave rise to it. It is an educative process in itself, whether or not one accepts their conclusions" (Lodge 1988:13). Keeping these theoretical underpinnings as a backdrop, a few literary narratives, Kamala Markandaya's *Nectar in a Sieve* (2009) and *The Coffer Dams* (2008), Shivram Karanth's *Return to Earth* (2002), Pundalik Naik's *The Upheaval* (2002), Indira Goswami's *The Man from Chinnamasta* (2006), Indra Sinha's *Animal's People* (2007), Sarah Joseph's *Gift in Green* (2011) and Na D'Souza's *Dweepa* (2013), are selected to analyse the ecofeminist discourse in Indian context. Therefore, the study deals with two important domains of enquiry: ecofeminism and the Indian novel.

Ecofeminism: environmental studies and feminism

Ecofeminism, a multidisciplinary intellectual and political movement that gained momentum in the 1980s, is an amalgamation of environmental studies and feminism. "Environmental studies," Jay Parini writes, "began in the sciences – Geology, Biology, Meteorology – but it has widened its embrace to include humanities and social sciences" (Parini 1995:15). It is, in a way, a meeting place of the environmental studies and feminism; the history of both the enterprises is quite interesting. Further, the interface, it seems, between nature and human expressions is ubiquitous across all cultures. However, the point is that the emergence of environmental studies as an academic discipline took place in the 1950s. The proliferation of scientific discourse on ecological problems gradually resulted in the birth of a new scientific discipline called Environmental Science. At the same time, commonsensical ideas regarding the protection of environment were also in circulation. The scientific discourses and the commonsensical ideas in society very subtly compelled the political interference too. Thus, we have the birth of an independent Ministry of Environment, formed in 1985. Feminists welcomed such a development and perhaps they were waiting for a kind of identity politics to be appropriated in the emergence of a new ecological concern.

The development of feminism coincided with the arrival of discourses on the ecological concern. The feminists obviously appropriated the politics and discursive formations of ecological concerns. The logic behind the identity politics of feminism was that man exploits both – woman and nature – and, hence, the common target of the discourse on nature and woman is patriarchy. Here, Noël Strugeon adds a further insight to our understanding of this type of identity politics, "Ecofeminism as a term indicates a double political intervention, of environmentalism into feminism and feminism into environmentalism" (Strugeon 1997:169). The feminists have found their common enemy in the patriarchal power structure.

The yoking together of environmental problems and feminism is quite an interesting intellectual enterprise that has occurred in the recent past. The two independent narratives – the narratives on environmental crisis and the narratives on feminism – joined hands to put forth their politics in the best sense of the term. The result has been ecofeminism. The ecofeminist discourse, which is a fusion of ecological studies and feminism, is a new discourse in the history of Humanities. The fusion of both these discourses has given birth to ecofeminism. It is an intellectual enterprise that has emerged in the late twentieth century. Its main agenda is to yoke feminine problems with environmental problems. It has been gradually evolving as a praxis-oriented theory and also considered as quilt theory: "it is structurally pluralistic, rather than structurally reductionist or unitary: it emerges from a multiplicity of voices, especially women's voices across cross-cultural context" (Warren 1994:84).

Ecofeminism, a neologism, was coined by Francoise d'Eaubonne, a French writer, in 1974 in her path-breaking work *Le Féminisme ou la mort*. Her fundamental intention was to give a call to women to save the planet. Further, it evolved as a theory by the contribution of various ecofeminists. Thirty-seven years ago, 2,000 women gathered at Pentagon in Washington DC, to protest military violence, with a consensual statement, "We have come to mourn and rage and defy the Pentagon because it is the workplace of the imperial power which threatens us all. They have accumulated over 30,000 nuclear bombs at the rate of three to six bombs every day. They are determined to produce the billion-dollar MX missile" (Paley 2008:461). This action was plausibly the consequence of the first women-led conference "Women and Life on Earth" held at Amherst in 1980. This protest, rather, was not just to consider ecological crisis but also to connect women and nature. This embryonic protest was a launch pad for various academic endeavours and activist movements to fight against the destruction of environment and exploitation of women across the globe. These women-led movements and protests also show that the concern of women towards nature is present in all cultures around the world. Women and ecology, thus, seem to be intertwined. These aspects have strongly influenced the birth of ecofeminism.

Let us move with this simple description of ecofeminism in this introductory chapter as a preliminary to formulate a line of inquiry. The second chapter, however, will give a detailed account of the discursive formation and theoretical postulates of ecofeminism.

Ecofeminism: Indian polemical discourse

In India, we have an abundant body of writings that deal with the problem of ecology. Writings on nature are the integral part of Indian literature. Along with the literary canon, India has also produced polemical discourse on environmental issues. Here before going to study Indian novel, we will have a cursory glance on the Indian polemical discourse. Sundarlal Bahuguna, a noted Garhwali Environmentalist, has written *Dharti Ki Pukar* (2007). His works mainly focus on the impact of dam construction on forest and water resources and the life of local people. He protested the construction of Tehri Dam. Ashis Nandy, Indian psychologist and social theorist, is also a critic of development, modernity, technology and science. Nandy's *Science, Hegemony and Violence: A Requiem for Modernity* (1988) – an anthology of essays – articulates the impact of two modern concepts such as science and development on nature and human beings. Nandy argues that these concepts demand sacrifices from ordinary citizens. Whereas P Sainath's *Everybody Loves a Good Drought* (1996) concentrates on myriad environmental issues related to land, water, forests, tribes and displacement. Amita Bavishkar's *In the Belly of the River* (1995) also addresses the impact of development on human beings and environment. Her concern in this work is to show the impact of the construction of the Sardar Sarover Dam on people of that area – the Bhilala tribal community of Anjavara. In addition to this, C K Janu, a social activist, is a leading light in tribal communities. Her autobiography, *Mother Forest: The Unfinished Story of C.K. Janu* (2004), acknowledges her constant struggle and strive to restore the land of the tribal people and their identity. Moreover, Anitha Agnihotri's *Forest Interludes* (2001) is kaleidoscopic view of fiction, memoir, personal essays and documentary, which are out of her experiences as IAS officer, of the poorest tribal areas of Orissa and east-central India. It reflects the stark reality Indian tribal people encounter with development and modernisation. If we throw a cursory glance at the above discussion, we will infer that in the Indian environmental crisis, both men and women are equally involved in the movement against modern developmental attitudes.

A few Indian ecofeminists have given their account of impact of ecological crisis on women and nature through their works. The seminal Indian ecofeminist is Vandana Shiva, whose major contribution to this field is through her works; *Ecofeminism* (1993), *Monoculture of the Mind: Perspectives on Biodiversity and Biotechnology* (1993), *Staying Alive: Women, Ecology and Survival in India* (2010), *The Violence of the Green Revolution* (1993) and *Biopiracy* (2012). Moreover, a few more Indian ecofeminists have formulated their own perspectives on the problem: for example, Bina Agarwal's "Environmental Management, Equity and Ecofeminism: Debating India's Experience" (1998), Chayya Datar's *Ecofeminism Revisited* (2011), Aruna Gnanadason's "Tradition of Prudence Lost: A Tragic World of Broken Relationship" (2003) and Manisha Rao's *Theory and Practice of Ecofeminism in India: An Analysis* (1996). The insights of these works will be discussed further in the second chapter. In the context of the pressing need for addressing the ecological crisis across the globe, it is also important to understand the insights offered by other nature writings, and in our case, we look to the novel.

Indian novel

Indian novels have been studied from a variety of perspectives, especially, largely from social (Bhatnagar 2001; Mukherjee 2002), political (Kaushik 1988, 2001; Prasad 2001; Bhatnagar 2007), historical (Naik 1982, 1985; Iyengar 1962), and cultural perspectives (Ganguli 1977; Lannoy 1971). In addition, regional novels have been studied and researched by the critics of their respective languages. However, these novels have hardly been studied from the perspective of ecocriticism and ecofeminism although the ecological crisis forms the backdrop of several Indian novels.

Some of the Indian novels address the contemporary environmental crisis, which has occupied a prominent place in the global humanities debate, and also insist us to have an ecological conscience. To mention a few Munshi Premchand's *Godan* (1936), Jim Corbett's *Man-Eater of Kumaon* (1944), Kamala Markandaya's *Nectar in a Sieve* (2009), Ruskin Bond's *The Blue Umbrella* (1974) and *The Cherry Tree* (1980), Rohinton Mistry's *Such a Long Journey* (1998), Gita Mehta's *A River Sutra* (1993), and Arundhati Roy's *The God of Small Things* (2012). Roy's four volumes of non-fiction writings, *The Algebra of Infinite Justice* (2001), *An Ordinary Person's Guide of Empire* (2005), *Listening to Grasshoppers* (2009), and *Broken Republic* (2011); Sohaila Abdulali's *The Madwoman of Jogare* (1999); Anita Nair's *Better Man* (2015); Mahasweta Devi's *The Book of the Hunter* (2002); Amitav Ghosh's *The Hungry Tide* (2004), *Sea of Poppies* (2008) and *River of Smoke* (2011); Anuradha Roy's *An Atlas of Impossible Longing* (2008); Mamang Dai's *The Legends of Pensam* (2006); Tamsula Ao's *These Hills Called Home Stories from a War Zone* (2006) and *Lamburunum for My Head Stories* (2009); Usha K.R's *Monkey-Man* (2010); and Ruchir Joshi's *The Last Jet-Engine Laugh* (2012), have all attempted to explicate the web of relationship among human beings, wildlife and nature.

To analyse a few among these, Arundhati Roy's, one of the most well-known Indian writers, *The God of Small Things* (2012) and her five volumes of non-fiction writings, *The Algebra of Infinite Justice* (2001), *An Ordinary Person's Guide of Empire* (2005), *Listening to Grasshoppers* (2009), *Broken Republic* (2011) and *Walking with the Comrades* (2011), project neoliberalistic perspective: the impact of globalisation, nuclear weapon, war, dam construction, etc., on human and nature. The major highlights of her essays are the problems of marginalised people and nature in the name of development and economic growth. Amitav Ghosh's *The Hungry Tide* (2004), *Sea of Poppies* (2008) and *River of Smoke* (2011) portray the anthropocentric attitude. *The Hungry Tide* discerns ambiguous and ineffective government policies and a clash between the Government and environmentalists with regard to the announcement of National Parks and Reserved Forests. Mahasweta Devi is a campaigner to tribal communities and radical social-political activist. Her works *Bitter Soil* (1998) and *The Book of the Hunter* (2002) are testimonies to the adverse effect of new settlements in the forest area on the lives of tribal communities. The socio-economic development not only disrupts the lives of the tribal people but also ruptures their socio-cultural norms.

Apart from the novels, the other literary genres exhibit the aspects of environment in their narration. The major nature-loving poets are from the Northeast region of India. Mamang Dai's 'River' and 'The Missing Links,' Dayananda Pathak's 'Coral Island,' Tamsula Ao's 'The Nightingale of Northeastern India' and so on are some of the examples. Along with these, a few mainstream poets have tried to show natural phenomenon and ecological erosion. For example, Toru Dutt's 'The Lotus' and 'Our Casuarina Tree,' S K Chettur's 'Red Lotus,' Rabindranath Tagore's 'Breezy April,' Sarojini Naidu's 'Summer Woods,' Dilip Chitre's 'The Felling of the Banyan Tree,' Gieve Patel's 'On Killing a Tree' and Keki Daruwalla's 'Boat-Ride Along the Ganga.

The Northeastern writers – Mamang Dai, Yeshe Dorji Thongchi (Arunachal Pradesh), Birendra Kumar Bhattacharyya, Hem Barua, Indira Goswami, Arup Kumar Dutta, Sanjib Baruah (Assam), Yumlembam Ibomcha (Manipur), Siddartha Deb, Anjum Hasan (Meghalaya), Mona Zote (Mizoram), Temsula Ao, Easterine Iralu, Charles Chasie, Anungla Aier (Nagaland) – deal with core issues of their region that is a confrontation between modernity and traditionality based on their regional ecology, that is, geopolitics. Mamang Dai, a prolific writer of Arunachal Pradesh, narrates a tribal lore in *The Legends of Pensam* (2006). It documents indigenous people lifestyle, whose base rooted into nature. For Arunachal Pradesh people, mountain, river and forest are not only the natural resources but also part and parcel of their lives. Most of their religious practices are related to mountain. The bottom line of these novels is the impact of social and economic development on tribal community, indigenous people and natural resources. The brief catalogue of the literary narration, a canon formulation for the selection of the novels for analysis, demonstrates the discursive formation of environment in Indian literary genre.

1. *The apropos of the select Indian novels*

This sketchy overview of a few Indian novels may raise a question that in spite of having ample literary narratives on ecological crisis in India. Why have I chosen the eight Indian novels for analysis? To answer this question: the first reason is that the most of Indian literary narratives represent Indian environmental crisis from ecocritical perspectives but the select Indian novels exhibit ecofeminist stance and this has set the foundation for the enquiry: In what way Indian novels add a new dimension to the ecofeminist discourse? The second reason is that these novels reflect the essence of Indian environmental problems – the contemporary reality. They are germinated from the soil of India and tinged with the novelists' personal experiential ethos. The dominant mode of the narration in these novels is realism. The task of the novelists is not only to document the facts and figures but also to recreate the real human experience artistically which paves a way to test the theory of ecofeminism in the Indian context.

A close examination of the Indian novels – based on the writers' empirical understanding of environmental crisis – reveals that these novels are many mouth pieces with single authorial voice narrating that ecological problems are general human problems. On the one hand these novels articulate the major concerns of the standard ecofeminist discourse such as the critique of patriarchal development attitude, women as saviours and sufferers of the environmental disaster, etc.; on the other hand, these novels narrate that it is appropriate to look at the ecological crises more as a general human problem than merely as a gender problem. The former gives us a partial picture of the idea of the ecological crisis whereas the latter depicts the panoramic vision of ecological crisis. The overarching emplotment of these novels represent shift from feminism to humanism. This paradigm shift is a cornerstone in the discursive formation of this enquiry.

To demonstrate this transition, the evaluation of the novels is primarily divided into two sections: the first section is the ecofeminist literary criticism and the second section manifests the uniqueness of the Indian novels and adds a new dimension to ecofeminist discourse. My viewpoint endorses that the study of any subject or discourse needs reassessment of that discourse on the basis of its specific context along with historical circumstances, as Edward Said argues, "that each humanistic investigation must formulates the nature of that connection in the specific context of the study, the subject matter, and its historical circumstances" (Said 2001:15). Therefore, the detailed analysis of Indian novels compels us to rethink the theoretical formulation of ecofeminism. That is, the environmental problems are not merely feminine problems; but, they are human problems because the ecofeminist discourse is the context-centric theory purely based on particular social, political and economic contexts.

Therefore, the present book proposes to reconsider ecofeminism as ecohumanism. Ecohumanism propounds that human beings – both women and men – and nature are part of a complex web. On the basis of the critical evaluation of the novels, I have framed a working definition of ecohumanism as, "an equivalence relation of man and woman by dissolving dualistic approach to resolve ecological crisis through common consensus." This book does not claim to have theorised this concept fully, but it has identified an important trait of ecological problems in India. I fervently feel to go beyond the humanities where human cannot be an exclusive category (Fukuyama 2002). To speak of the human is to speak of a home located in a particular place of which the human is part (Selvamony 2014).

2. *The strategy to bring the select novels together*

Although literary historians raise the problem of tradition while conceptualising regional Indian novels belonging to a single tradition, the study does not make a distinction between the Indian English novel and Indian regional novels. Since the focus of the study is thematic and deals with the category of a particular nation, India, the question of these novels belonging to a single tradition is not paid much attention to. However, as the aim of this study is to locate Indian writers' response to the crisis of ecology and feminism, the selection of Indian novels from different traditions will not hamper our understanding of the issue under consideration and will not come in our way of forwarding our hypothesis and building our arguments. Therefore, all Indian novels irrespective of their medium of expressions are treated as belonging to the Indian tradition, especially to do critical investigation of the ecofeminist discourse in Indian context, because of their single coherent voice. In case of novels written in Modern Indian languages (other than English), the widely accepted translations of such novels into English are used for the analysis here.

The select Indian novels are classified on the basis of their treatment of thematic concern: Kamala Markandaya's *Nectar in a Sieve* (2009), Pundalik Naik's *The Upheaval* (2002), Shivram Karanth's, *Return to Earth* (2002) and Sarah Joseph's *Gift in Green* (2011) narrate the impact of tannery, mining, modernisation, globalisation and industrialisation on agriculture. Therefore, they are clubbed in the third chapter under the framework of agriculture and ecofeminism. Then, Kamala Markandaya's *The Coffer Dams* (2008) and Na D'Souza's *Dweepa* (2013) narrate the impact of dam construction on native people and nature. For this reason, they are grouped together under the category of dam construction as leading to ecological crisis. Indira Sinha's *Animal's People* (2007) draws our attention to the effect of industrialisation on nature and human beings. Hence, it is categorised under the framework of industrialisation leading to ecological problem. Indira Goswami's *The Man from Chinnamasta* (2006) looks at animal sacrifice as an ecological crisis. Therefore, the novel is discussed under the category of Animal's as absent category.

3. *A conglomeration of various disciplines*

To address the contemporary environmental crisis, many literary genres and studies have been introduced in literary academe. All these ecological discourses – literary narratives or polemical discourses – cannot be studied in the isolation because they are interdependent. The book's focus, ecofeminist reading of Indian novels, opens frontiers for many disciplines and discourses especially for the academicians and students of literature and criticism; it has layered perspectives for interpretations because it is not only limited to literary narratives but it probes into various issues related to ecological crisis which forms the backdrop for the select Indian novels: the detailed accounts of mining in Goa and the landfilling of hundreds of acres of water-beds and mangrove forests in Ernakulam district to build a cricket stadium which impact on Indian agrarian culture, Indian project of dam construction and its effect on environment, industrial disaster of Bhopal, the Indian mythological background and the ethnography of the animal sacrifice, etc. Along with these factual illustrations, it offers various instances of the interrelations of social, economical, political and cultural issues – related to ecological crisis – which are the sights for the study to the research scholars and the academicians of environmental humanities, ecological humanities, Green Studies and Literature, ecocriticism, ecofeminism, anthropologists, environmentalists, environmental justice, political ecology, etc.

Bibliography

Abulali, Sohaila. *The Madwoman of Jogare*. HarperColins, 1999.

Agarwal, Bina. "Environmental Management, Equity and Ecofeminism: Debating India's Experience." *Journal of Peasant Studies*, vol.25, no.4, 1998, pp. 55–95.

———. "The Gender and Environment Debate: Lessons from India." *Feminist Studies*, vol.18, no.1, 1992, pp. 119–158.

Agnihotri, Anitha. *Forest Interludes*. Translated and edited by Kalpana Bardhan. Kali for Women, 2001.

Ao, Temsula. *Laburunum for My Head*. Penguin Books, 2009.

———. *These Hills Called Home Stories from a War Zone*. Zubaan-Penguin Books, 2006.

Bahuguna, Sundarlal. *Dharti Ki Pukar*. Radhakrishan Prakashan, 2007.

———. *Walking with the Chipko Message* (Teheri Garhwal District). Navijivan Asharam, 1983.

Baviskar, Amita. *The Belly of the River: Tribal Conflicts over Development in the Narmada Valley*. Oxford University Press, 1995.

Bhatnagar, Manmohan. *Indian Writing in English Volume II*. Atlantic Publishers, 2001.

Bhatnagar, O.P. *Indian Political Novel in English*. Sarup & Sons, 2007.

Biehl, Janet. *Rethinking Ecofeminist Politics*. South End Books, 1991.

Birkeland, Janis. "Ecofeminism: Linking Theory and Practice." *Ecofeminism: Women, Animals, Nature*, edited by Greta Gaard. Temple University Press, 1993.

Bond, Ruskin. *Cherry Tree*. Boyds Mills Press, 1996.

———. *The Blue Umbrells*. Red Turtle, 2013.

Booth, Annie. "Does the Spirit Move You? Environmental Spirituality." *Environmental Values*, vol.8, no.1, 1999, pp. 89–105.

Buege, J. Douglas. "Rethinking Again: A Defense of Ecofeminist Philosophy." *Ecological Feminism*, edited by Karen Warren. Routledge, 1994.

Caldecott, Leonie, and Stephaine Leland, editors. *Reclaim the Earth: Women Speak Out for Life on Earth*. Women's Press, 1983.

Cather, Willa. *O Pioneers*. Hesperus Press, 2013.

Corbett, Jim. *Man-Eaters of Kumaon*. Oxford University Press, 1989.

Dai, Mamang. *The Legends of Pensam*. Penguin Books, 2006.

Datar, Chhaya. *Ecofeminism Revisited*. Rawat Publications, 2011.

D'Eaubonne, Francoise. "Feminism-Ecology: Revolution or Mutation?" *Ethics and the Environment*, vol.4, no.2, 1999, pp. 175–177.

———. *Le feminism ou la mort*. Pierre Horay, 1974.

Devi, Mahasweta. *Bitter Soil*. Seagull Books, 1998.

———. *The Book of the Hunter*. Seagull Books, 2002.

D'Souza, Na. *Dweepa*. Translated by Susheela Punitha. Oxford University Press, 2013.

Fukuyama, Francis. *Our Posthuman Future: Consequences of the Biotechnological Revolution*. Profile Books, 2002.

Ganguli, B.N., editor. *Tradition, Modernity and Development: A Study in Contemporary Indian Society*. Macmillan, 1977.

Ghosh, Amitav. *The Hungry Tide*. HarperCollins, 2004.

———. *River of Smoke*. Picador, 2011.

———. *Sea of Poppies*. John Murray, 2008.

Gnanadason, Aruna. "Traditions of Prudence Lost: A Tragic World of Broken Relationships." *Ecofeminism and Globalization*, edited by Heather Eaton and Lois Ann Lorentzen. Rowman & Littlefield Press, 2003, pp. 73–87.

Goswami, Indira. *The Man from Chinnamasta*. Translated by Prashant Goswami. Katha, 2006.

Gundur, N.S. *Varthamanada Itihasakara: Michael Foucault*. Kuvempu Bhasha Bharati Pradhikara, 2017.

Iyengar, Srinivasa. *Indian Writing in English*. Asia Publishing House, 1962.

———. "On Re-Reading the Serpent and the Rope." *Perspectives on Indian Fiction in English*, edited by M.K. Naik. Abhinav Publications, 1985.

Janu, C.K. *Mother Forest: The Unfinished Story of C. K. Janu*. Translated by N. Ravishankar. Kali for Women, 2004.

Joseph, Sarah. *Gift in Green*. Translated by Valson Thampu. Harper Perennial, 2011.

Joshi, Ruchir. *Jet-Engine Laugh*. HarperCollins Publishers, 2012.

Karanth, Shivram. *Return to Earth*. Translated by Padma Ramachandra Sharma. Sahitya Akademi Centre for Translation, 2002.

Kaushik, Asha. *Political Aesthetics and Culture: A Study of Indo-Anglican Political Novel*. Manohar, 1988.

———. *Politics, Symbols, and Political Theory: Rethinking Gandhi*. Rawat, 2001.

Lannoy, Richard. *The Speaking Tree: A Study of Indian Culture and Society*. Oxford University Press, 1971.

Lodge, David, editor. *Modern Criticism and Theory*. Pearson Education Ltd, 1988.

Markandaya, Kamala. *The Coffer Dams*. Penguin Books, 2008.

———. *Nectar in a Sieve*. Penguin Books, 2009.

Mehta, Gita. *A River Sutra*. Vintage, 1994.

Mies, Maria, and Vandana Shiva. *Ecofeminism*. Rawat Publications, 1993.

Mistry, Rohinton. *Such a Ling Journey*. Mcclelland & Stewart, 1998.

Mukherjee, Meenakshi. *Early Novels in India*. Sahitya Akademi, 2002.

———. *Realism and Reality: The Novel and Society in India*. Oxford University Press, 1985.

———. *The Twice Born Fiction*. Arnold-Heinemann Publishers, 2016.

Naik, M.K. *A History of Indian English Literature*. Sahitya Akademi, 1982.

———, editor. *Perspectives on Indian Fiction in English*. Abhinav Publications, 1985.

Naik, Pundalik. *The Upheaval*. Translated by Vidya Pai. Oxford University Press, 2002.

Nair, Anita. *The Better Man*. Penguin, 2015.

Nandy, Ashis, editor. *Science, Hegemony and Violence: A Requiem for Modernity*. Oxford University Press, 1988.

Paley, Grace. "Women Pentagon Action Unity Statement." *The Massachusetts Review*, vol.49, no.4, 2008, pp. 461–464.

Parini, Jay. "The Greening of the Humanities." *The New York Times Magazine*, 29 October, 1995, p. 15.

Pollock, Sheldon. *The Language of the Gods in the World of Men*. Permanent Black, 2006.

Prasad, Amarnath. *Critical Response to Indian Fiction in English*. Atlantic Publishers, 2001.

Premchand, Munshi. *Godan*. Vishv Vijay Pvt Ltd, 1936.

Rao, Manisha. "Ecofeminism at the Crossroads in India: A Review." *DEP*, vol.20, 2012, pp. 124–142.

———. *Theory and Practice of Ecofeminism in India: An Analysis*. Women's Studies Centre, Department of Sociology, University of Pune, 1996.

Roy, Arundhati. *The Algebra of Infinite Justice*. Penguin India, 2001.

———. *Broken Republic*. Hamish Hamilton, 2011.

———. *The God of Small Things*. Penguin Books, 2012.

———. *Listening to Grasshoppers*. Hamish Hamilton, 2009.

———. *An Ordinary Person's Guide of Empire*. Penguin India, 2005.

———. *Walking with the Comrades*. Penguin Books, 2011.

Said, Edward. *Orientalism*. Vintage Books, 2001.

Sainath, P. *Everybody Loves a Good Drought*. Penguin Books, 1996.

Selvamony, Nirmal. "Considering the Humanities Ecotheoretically." *Journal of Contemporary Thought*, vol.40, 2014, pp. 5–19.

Selvamony, Nirmal, et al. *Essays in Ecocriticism*. Sarup & Sons and OSLE-India, 2008.

Shiva, Vandana. *Biopiracy*. Natraj Publishers, 2012.

———. *Monoculture of the Mind: Perspectives on Biodiversity and Biotechnology*. Zed Books Ltd, 1993.

———. *Staying Alive Women, Ecology and Survival in India*. Women Unlimited, 2010.

———. *The Violence of the Green Revolution*. Zed Books Publication, 1993.

Sinha, Indra. *Animal's People*. Simon & Schuster, 2007.

Strugeon, Noël. *Ecofeminism Natures: Race, Gender, Feminist Theory and Political Action*. Routledge, 1997.

Usha, K.R. *Monkey-man*. Penguin, 2010.

Warren, J. Karen. *Ecological Feminism*. Routledge, 1994.

1 An interface between human beings and nature

1.1 Introduction

The aim of the book, as stated in the introductory chapter, is to examine the ecofeminist discourse in Indian context through the Indian literary narratives. To pursue this aim, it is very crucial to understand the theory of ecofeminism, its historical evolution and basic tenets. Therefore, this chapter is mainly to formulate a theoretical framework to read Indian novels by giving an illuminating sketch of the theory of ecofeminism as postulated by seminal ecofeminists. To this end, this chapter presents the genealogy of ecofeminism, the genealogy of the Indian novel, the discursive formation of the environmental narration in Indian novels and in polemical discourse.

1.2 The genealogy of ecofeminism

In order to have a comprehensive understanding of ecofeminism, we need to understand the term 'ecology,' which has been derived from two Greek words *oikos* and *logos; oikos* means 'house' or 'place to live in' and *logos* means 'study' (Haeckel 1866). Therefore, ecology is a study of interrelationship between all living organisms and their environment. Reiter coined the term 'ecology' in 1868. The traditional definition of ecology is "the study of reciprocal relationships between organisms and their environment." Ernest Haeckel (1866), a German biologist, introduced the meaning of 'ecology' defined as: "the total relation of the animal to both its organic and its inorganic environment." Further, Eugene Odum (1953, 1996) expounds it as "the study of the structure and function of ecosystems or structure and function of nature." The Indian ecologist R. Mishra's (1967) definition of ecology is "interactions of forms, functions and factors."

The reciprocal relationship between human beings and their environment is gradually disrupted by consumerism, industrialisation and scientific revolution, which lead to environmental crisis. This crisis has given way to many studies, philosophies and theories to solve environmental problems. In 1972, Arne Nass, a Norwegian philosopher, coined two terms: 'deep ecology' and 'ecosophy', which have been introduced to environmental literature. Deep ecology is an environmental philosophy and a movement that states that human beings and non-human beings have intrinsic value and we are not able to separate human beings from natural environment. Ecosophy is a philosophy of ecological harmony and equilibrium within human beings and nature. Along with these two concepts, a few more terms have been introduced to environmental literature; one, Shallow Ecology, which is anthropocentric and considers nature as instrumental, and two, Ecological Marxism, which emphasises the need to restructure modern production conditions. New Positional Ecology focuses on "locality-based studies of people insertion with their environments" (Goldman & Schurman 2000:568). Next, Theodore Roszak used a new neologism ecopsychology, in his book *The Voice of the Earth* (1993), which tries to locate the relationship between human beings and the natural world through ecological and psychological principles.

Further, we also need to pay attention to the development of a critical practice called Ecocriticism. The term ecocriticism was first used by William Rueckert (1978) in his article 'Literature and Ecology: An Experiment in Ecocriticism' in order to create an awareness of environmental crisis through literary genre. Cheryll Glotfelty and Harold Fromm's book *The Eco-Criticism Reader: Landmarks in Literary Ecology* (1996) played an instrumental role in accepting ecocriticism as a literary theory. Cheryll Glotfelty, in his introduction to the book, defines ecocriticism as 'the study of the relationship between literature and the physical environment' (xviii). Another seminal writer, Lawrence Buell has contributed to the growth of ecocriticism as a literary discipline through his book, *Writing for Endangered World* (2001). His work, *The Environmental Imagination: Thoreau, Nature Writing and the Formation of American Culture* (1995), highlights Henry David Thoreau's concern for nature. Thoreau's writings reflect on nature-centric life; he was in search of an alternative natural mode of living to replace modern lifestyle. The Japanese agriculturalist and philosopher, Masanobu Fukuoka's, *One Straw Revolution* (1975), a book on natural farming, attempts to show the link among literature, philosophy, agriculture and environmental issues. Rachel Carson's epoch-making book *Silent Spring* (1963) is a critical introduction to the field of ecocriticism. It discusses the impact of pesticides on birds and nature. Bate's more recent book *The Song of the Earth* (2000) argues that colonialism and deforestation have frequently gone together for the destruction of nature at its extent (Barry 162). Further, the most important contribution to this field is by Greg Garrard, who in his seminal work *Ecocriticism* (2004), says ecocriticism means 'ecological literacy.' He very aptly defines, "Ecocriticism is the study of the relationship of the human and non-human, throughout human cultural history and entailing critical analysis of the term 'human' itself." He further espouses, "Developing the insights of earlier critical movements, ecofeminists, social ecologists and environmental justice advocates seek a synthesis of environmental and social concerns" (Garrard 2004:3, 5).

Even Richard Kerridge's definition of ecocriticism, in his monumental work *Writing the Environment* (1998), is parallel to Cheryll Glotfelty's definition, "The ecocritic wants to track environmental ideas and representations wherever they appear, to see more clearly a debate which seems to be taking place, often part concealed, in a great many cultural spaces. Most of all, ecocriticism seeks to evaluate texts and ideas in terms of their coherence and usefulness as responses to environmental crisis" (Kerridge 1998:5).

Ecological crises has given birth to a seminal theory called as ecofeminism. Ecofeminism, one of the upcoming monumental movements, addresses the exploitation of women and nature. Ecofeminism, an umbrella term, focuses on the philosophical, practical and analytical study of exploitation of women and nature. It is a new approach to feminism and environmentalism. Ecofeminism is a study of exploitation of both women and nature and their liberation. Ecofeminism is a portmanteau – ecology and feminism. Ecology refers to the scientific study of living organisms and their relationship with each other in the natural environment. It is concerned with the study of the effect of civilisation on environment and the need to conserve nature by preventing its exploitation by human beings (Mishra 2008). On the other hand, feminist theories have emerged from feminist movements and their basic purpose is to understand gender inequality and to focus on gender politics, power relations and sexuality (Abrams 2009). Thus, feminists study reasons for exploitation and marginalisation of women and environmentalists study reasons for the exploitation of nature. In their endeavour to find answers to their question – who or what is the root cause of the exploitation of woman and nature? – ecofeminists find their answer in patriarchal development attitude. Therefore, the common enemy for ecocritics and feminists is the powerful gender, the man. The ultimate goal of both these movements is to liberate their respective subjects, that is, women and nature from the patriarchal power relationship. It is not just an intellectual enterprise. It is through and through a socio-political movement. It has stemmed out of various academic endeavours and activist movements against the destruction of environment across the globe with the diagnostic focus on women and nature. Ecofeminism, as a movement, is an outcome of the ecological crisis brought about by industrialisation, modernisation and the growth of market culture.

The concern of women towards ecology and nature has been noted in all cultures around the world. Women and ecology, thus, seem to be intertwined. These aspects have strongly influenced the birth of ecofeminism. Ecofeminism was coined by French writer Francoise d'Eaubonne in 1974 in her path-breaking work *Le Féminisme ou la mort*. Her fundamental intention was to give a call to women to save the planet. Further, ecofeminism has been established as a movement, where women are protesters, held academic activities and activist movements against environmental destruction. Ynestra King, an ecofeminist theorist, developed the concept further by organising a conference. The melting down of Three Mile Island alerted a number of women in the USA to contemplate the situation and gather for the first ecofeminist conference entitled "Women and Life on Earth: A Conference on Ecofeminism in the Eighties," in March 1980, at Amherst. The other significant conferences on ecofeminism are "Ecofeminist Perspectives: Culture, Nature, Theory" in March 1987 at the University of Southern California, the international political conferences sponsored by the UN and international NGOs such as the conference on women "UN's Environment Programme (UNEP) Global Assembly and the Environment," which was held in Nairobi in 1985. And the "World Women's Congress for a Healthy Planet," held in 1991 in Miami, not only brought ecofeminists on one platform but also focused exclusively on environmental and women issues (Hobgood-Oster 2005).

In addition to these conferences, numerous protests by women against ecological destruction and deterioration were also held. The notable among such protests are: the first women protest action was Women's Pentagon Action in United States (Carlassare 1994:220), the peasant women protest against the proposed construction of the nuclear power plant at Whyl in South West Germany, the protest against chalk mining and logging in the Himalayas that is Chipko movement, the Green Belt Movement in Kenya, the protest by the Japanese women against chemically stimulated food and commercial agriculture asking for self-reliant producer-consumer networks, poor women's efforts in Ecuador to save the mangrove forests as breeding-grounds for fish and shrimp, the battle of thousands of women in the South for better water management, soil conservation, land use and maintenance of their survival base against the industrial interests, Medha Patkar's fight against the construction of Narmada Dam, Lois Gibbs' opposition to the dumping of toxic waste in Love Canal, Joan Sharp's struggle to close the Schlage Lock Company in North Carolina, United States, which was responsible for the death of 30 women due to cancer and the protest by Russian women against the Chernobyl catastrophe (Mies & Shiva 4). Another example of women-led protest is women's peace camp against operating and testing nuclear cruise missiles at Greenham common in England (Gaard 2011:29). These protests to save the planet for future generations have provided a good platform for ecofeminism to gain currency as a movement. These socio-economic movements do not occur without the support of ideas. Ideas and ideals have changed this world. Behind the movement of ecofeminism, there is a strong foundation of polemical thought on ecological and feministic problems. To analyse Indian novels from the perspective of ecofeminism, we need to understand polemical intricacies of the ecofeminist discourse. The disparity compels us to do a systematic study of the major ecofeminists' perspectives in order to get a coherent general framework. Later, many ecofeminists added their perspectives and concerns to ecofeminism to establish it as a theory.

This dual theoretical enterprise draws our attention to the patriarchal violence done to both women and nature in the name of development of science and technology, modernisation and globalisation. Thus, ecofeminism has been emphasising the exploitation of women and nature on the basis of a subtle connection between the two. Therefore, let us examine the close association and the subtle connection between women and nature.

The term ecofeminism may raise a question: how are the two very different disciplines ecology and feminism brought together under one term, ecofeminism? What exactly is the connection between women and nature? To answer these questions, it is quite necessary to understand the logic behind the creation of this portmanteau term. Let us examine a theoretical perspective that makes three claims – empirical, conceptual and epistemological – in order to show the connection between women and nature (Eaton & Lorentzen 2003; Warren 1996).

The first empirical claim shows that the first-hand victims of the impact of environmental deterioration are women. In this respect, Heather Eaton observes thus:

> Ecofeminists' empirical claim examines sociopolitical and economic structures that restrict many women's lives to poverty, ecological deprivation, and economic powerlessness. The degradation of environment has affected women in the most parts of the world. The United Nations in 1989 remarked, it is a now universally established fact that it is the woman who is the worst victim of environmental destruction. The poorer she is the greater is her burden.
>
> (Eaton and Lorentzen 2003:2)

Women, especially from the Third World and the Eastern countries, exhibit first-hand experience of exploitation of nature. In these countries, women are nurturers, caretakers and life givers. They are involved in domestic work on account of which they are not only very close to nature but also depend on nature for fuel, food, fodder and water. First, deforestation creates a direct problem for women, especially in the agrarian setting, as they have to travel long distances to collect fuel. In this context, it is very apt to study Ruether's line of thought. According to Ruether, "Deforestation means women walk twice and three times as long each day gathering wood; it means drought which means woman walk twice and three times farther each day to find and carry water to their modest houses" (Ruether 1997:40). Second, a woman labourer is not recognised equal to men and women are paid less when compared to men. Third, commercial agriculture has replaced the indigenous knowledge of seeds. Most of the women are involved in agricultural practices, but due to the introduction of corporate agriculture and new technology, they have lost their livelihood. As Vandana Shiva observes, "Peasants and farmers are thus robbed of their means of livelihood by the new technology which becomes an instrument of poverty and underdevelopment" (Mies and Shiva 1993:29).

Synthetic chemicals, pesticides and insecticides have adverse impact on women and nature. Added to this, Rachel Carson (1963), a scientist, expostulates that the poison of synthetic chemicals, pesticides and insecticides are being passed from mother to offspring. Food and Drug Administration scientists have found that residues of insecticides are present in mother's milk, and it is also proved that breastfed infant receives small amounts of toxic chemicals regularly. Even animal milk is contaminated with insecticides and pesticides (Carson 1963:31).

The second conceptual claim shows that patriarchal ideologies are the root causes of domination over women and exploitation of nature. Karen Warren aptly espouses as follows:

> A conceptual framework is a set of basic beliefs, values, attitudes, and assumptions which shape and reflect how one views oneself and one's world. It is a socially constructed lens through which we perceive ourselves and others. It is affected by such factors as gender, race, class, age, affection, orientation, nationality, and religious background. Some conceptual frameworks are oppressive. An oppressive conceptual framework is one that explains, justifies, and maintains relationships of domination and subordination. When an oppressive conceptual framework is patriarchal, it explains, justifies, and maintains the subordination of women by men.
>
> (Warren & Cady 1994)

The above conceptual claim means that the construction of society is on the basis of "hierarchy and dualism" (Eaton & Lorentzen 2003:2). This shows patriarchal ideologies are the root cause of domination over women and the exploitation of nature. There are three noteworthy features underlying this conceptual framework: "dualism" (Plumwood 1993, 2002), "value-hierarchical thinking" (Plumwood 1993) and "logic of domination" (Warren 1997).

Dualism not only depicts a relationship of difference, dichotomy and non-identity but also systematically constructs an inferior one. It has not only construed human identity and the relationship between human beings and nature but also designed Western views about nature. Dualism is a key concept in the Western thought which reflects major forms of oppression in the Western culture. Val Plumwood examines how the Western culture has treated the human/nature relationship in terms of dualism and how dualism has given a platform for many ecological problems. This dualistic Western approach to nature is a grassroots environmental crisis. At par, a disjunctive pair, men/women, gives a picture of superiority of men and inferiority of women and human/nature same as the first pair (Plumwood 1993:2).

The second significant feature of this conceptual claim is the value of hierarchical thinking, which means "up-down" thinking. Up refers to higher value, status and prestige, whereas down points towards lower value, status and prestige. Historically, women, nature and animals are considered as a down category. So, the up category has always exploited the down. In this context, Plato's cosmology is explained as a rational male form considered as cosmos and an irrational female matter considered as chaos. Plato's devaluation of women is very clearly projected in *Phaedo*, where Plato's debasement of women runs much deeper than mere personal distaste for women as a sex (Plumwood 1993).

The third feature of the conceptual claim is the 'logic of domination' that pertains to the domination of the superior over the inferior. In this context,

Bookchin asserts a historical reduction thesis, "All our notions of dominating nature stem from the very real domination of human by human. . . . As a historical statement [this] declares in no uncertain terms that the domination of human by human *preceded* the notion of dominating nature" (Bookchin 1989:44).

The third claim of connection between women and nature is an epistemological claim. Women are agrarian cultivators and their major focus is on sustainable and renewable agriculture. As Vandana Shiva has given the Food and Agriculture Organization's statistic report of Women Feed the World:

> Women use more plant diversity, both cultivated and uncultivated, than agriculture scientists know about. In Nigerian home gardens, women plant 18–57 plant species; in sub-Saharan Africa they cultivate as many as 120 different plants in the species left, alongside the cash crops managed by men; in Guatemala, home gardens of less than 0.1 hectare have more than ten tree and crop species. In a single African home garden, there are more than ten trees and crop species. In a single African home garden more than 60 species of food-producing trees were counted. In Thailand, researchers found 230 species in home gardens. In Indian agriculture, women use 150 different species of plants for vegetables, fodder and healthcare. In the Expana region of Veracruz, Mexico, peasants utilize about 435 wild plant and animal species of which 229 are eaten. Women are the biodiversity experts of the world.
>
> (Shiva 2010:x–xi)

For ages, women have contributed to the conservation of resources and soil fertility. But industrial agriculture, biotechnology, genetic engineering, hybridisation and global corporate market have replaced their primitive knowledge of agriculture. Shiva says, "Navadanya's studies on biodiversity-based ecological agriculture indicate that women run farms and produce more food and nutrition than industrial, chemical farms" (Shiva 2010:xi). Women support subsistence agriculture by using and managing biomass for food, fodder, fertilizer, and fuel, which retains soil fertility and ensures nature-friendly activities. Unfortunately, women remain invisible in food production; their work is unrecognised and often not measured in terms of wages.

Obviously, ecofeminism has emerged as an interdisciplinary endeavour by coalescing various issues related to ecology and women, "Almost all of them concerning the construction of the relationship between women and nature that lie at the core of this particular feminist ecological project" (Sandilands 1999:4). However, based on the seminal ecofeminists perspectives, the entire three claims outlined above legitimises three basic tenets: the first, women are exploited by patriarchal development attitude because of their close association with nature and dependency on it. Second, women are saviours of nature. Lastly, women are life givers, nurtures and caretakers of nature.

1.3 Indian polemical discourse: environment and ecofeminism

In this section, we are not only going to make a brief overview of Indian ecofeminists' standpoints but also the perspectives of the Indian environmentalists, philosophers and thinkers in order to understand the root cause of environmental problems in India.

1.3.1 The personal dimension

The nexus of my personal experience and knowledge of traditional practices, beliefs and customs of India establish that how nature is deep rooted in Indian culture. The concern for nature is not alien to Indian society and it goes way back to Indus valley civilisation. Nature was part and parcel of their lives. The Indus Valley People were worshipper of nature and animal. They worshipped animals like elephant, lion, bull, buffalo, tiger, unicorn, crocodile, dove and snake. The snake worshippers were considered as the Naga worshippers. All these animals were represented on seal and amulets or terracotta, faience and stone figures, some of them depicted semi human, half animal and half human beings. Some of them were considered mythical animals. This animal worship is practiced even today. Some animals have been associated with gods and goddess as their vehicles, such as Shiva's vehicle is the Nandi or Bull, Durga's is tiger, Vishnu's is Garuda and Ganesha's is rat and Krishna is associated with the cow. *Puranas* narrate many stories, myths and legends associated with cow. Hindus even today regard cow as a sacred animal. The sun is worshipped as God and many sun temples are constructed in India. The existence of tree worship in Indus Valley is evidenced by representation of trees and leaves, especially Pipal, on many seals and sealing. A large number of plants and trees like Tulsi, Pipal, Banyan, Neem, etc., are worshipped even today. Leaves of mango tree and Ashoka tree are used for many religious and social occasions (Luniya 1982 & Basham 1975).

Watering a Tulsi plant daily is a religious duty in many Hindu families. Water played a very vital and scared place in the religious belief of Indus Valley people. Bathing in sacred tanks and rivers like the Ganga, the Narmada, the Godawari, is considered as holy even today. The people of early Vedic age or Aryans were deeply impressed by nature and its mysterious workings, considered them as divine or supernatural characters. They realised the dependence of human welfare on the power of nature. They worshipped the sun, the moon, the sky, thunder, wind, air, etc. In spite of this vast cultural legacy, the environment is gradually degrading because of the development attitude and demand of the growing population.

1.3.2 Indian environmentalist view

Many Indian environmentalists have constantly striven to save nature – forest, water bodies and cultivated land from modern developmental projects. Let us have a glance at not a comprehensive catalogue of the environmentalists and writers but a sketchy view. Chandi Prasad Bhatt, a Gandhian environmentalist and activist, is the founder of *Dasholi Gram Swarajya Sangh* (Dasholi Society for Village Self-Rule). It is the mother organisation of the Chipko movement and aims to set up small industries using forest resources. Sunita Narian, an Indian environmentalist, is a promoter of the Green concept of sustainable development. M. S. Swaminathan is known as the father of the Green revolution in India; he is a promoter of sustainable development on the basis of sustainable agriculture, sustainable food security and the preservation of biodiversity. He is also the supporter of the movement – Save Silent Valley. Sugathi Kumari was a founder of *Prakrati Samrakshana Samithi* (Save Nature Society), and she too played a vital role in the movement of the Save Silent valley. Sundarlal Bahuguna is a leader of Chipko movement. Medha Patkar is a social activist who fights for the rights of adivasis, farmers and women. She is a crusader of the *Narmada Bachao Andolan* (Save the Narmada River Movement). Shehla Masood worked for wildlife conservation, Parasu Ram Mishra was an Indian soil conservationist, TGK Menon introduced environmental friendly irrigation and farming techniques and promoted biodynamic agriculture and Ranjit Bhargava strived to obtain UNESCO world heritage site status for the upper Ganges. These environmentalists constitute only the tip of iceberg of the vast fraternity of Indian environmentalists.

Apart from these environmentalists, many Indian thinkers and writers who are sensitive to environmental problems have put forth their perspectives in their works. We will assess the contribution of a few Indian thinkers and writers to this domain. Sundarlal Bahuguna, a noted Garhwali Environmentalist, has written *Dharti Ki Pukar* (2007). His work focuses mainly on the impact of dam construction on forest, water resources and the life of local people. He protested the construction of Tehri Dam. Ashis Nandy, Indian psychologist and social theorist, is another critic of development, modernity, technology and science. Nandy's *Science, Hegemony and Violence: A Requiem for Modernity* (1988) – an anthology of essays – articulates the impact of two modern concepts such as science and development on nature and human beings. Nandy argues that these concepts demand sacrifices from ordinary citizens.

Nirmal Selvamony, the head in charge of the Department of English Studies at Central University of Tamil Nadu, is the founder of a forum called *tiNai* (*oikos*). The meaning of *tiNai* is traditional organic relationship of human beings with nature (Selvamony et al. 2008; Selvamony 2014). Selvamony (2014) says that dualism and anthropocentrism are the root causes of the ecological crisis. P Sainath's *Everybody Loves a Good Drought* (1996) concentrates on myriad environmental issues related to land, water, forests, tribes and displacement. Amita Bavishkar's *In the Belly of the River* (1995) addresses the impact of development on human beings and environment. Her concern in this work is to show the impact of the construction of the Sardar Sarover Dam on people of that area – the Bhilala tribal community of Anjavara. If we throw a cursory glance at the above discussion, we will infer that in the Indian environmental crisis, both men and women are equally involved in the movement against modern developmental attitudes.

1.3.3 Indian ecofeminist standpoint

Now, we have to proceed to contemplate over a few Indian ecofeminists' contributions to the theory. Vandana Shiva, a prominent Indian ecofeminist thinker and activist, has made major contributions to the field of ecofeminism through her works, viz., *Ecofeminism* (1993), *The Violence of the Green Revolution* (1993), *Staying Alive: Women, Ecology and Survival in India* (2010) and *Biopiracy* (2012). According to Shiva, women and nature have an intricate and intimate relationship as well as a shared history on the grounds of exploitation, degradation and domination by the androcentric attitude. Shiva considers the emergence of a new world order – development, modernisation and advancements in science and technology – as the root cause of the exploitation of women and nature. She claims that women's subsistence practices, knowledge of nature and dependency on nature for staying alive have been marginalised and replaced with modern science and related practices. Due to this, indigenous people are physically and culturally uprooted from their ancestors' soil. Moreover, it has caused an onslaught on local cultures resulting in a fragmentation and commodification of local cultures into saleable entities at global super market in the form of 'ethnic food,' 'ethnic music' and 'folklore' by harnessing ethnic objects for the tourist industry.

Further, Shiva attacks such a development as a Western paradigm that has vicious effects on human beings in general and women and nature in particular. The basic purpose of such a development is economic gain, which is a key driving factor of the Western industrialised economy. The terms 'development', 'progress' and 'globalisation' appear to be synonyms for the Western countries. The Western countries are more focused on productivity and economic growth without giving much attention to what happens to nature. They assume that natural resources are available in abundance for satisfying their greed, which may not be limited by their needs. This attitude leads to destruction of *'Prakrati'* – an active, powerful and productive force essential for the renewal and sustenance of all life. Due to the exploitation of nature, women are getting deprived of their activity, creativity and sanctity of life. They are struggling to conserve their subsistence base. In this context, Vandana Shiva gives the example of the first epoch-making movement, the Chipko movement, in which the women of Garhwal district in Himalayas protested against the commercial felling of trees by hugging the trees.

Next, research institutions and the introduction of new technologies in agriculture have deprived peasants and farmers of their traditional agricultural practices. Especially in the case of the Third World, women peasants, who are producing and reproducing wealth in partnership with nature by maintaining the integrity of ecosystem, have been affected by the advent of new technology. Besides, reductionist forestry has destroyed tropical forest, while reductionist agriculture has destroyed tropical farming. A patriarchal mode of progress emphasises monoculture, in which a homogeneous system yields high productivity but only in one dimension based on commercial interests and economic imperative, whereas multi-cropping has lower productivity but is environmentally conservative and therefore in real terms it is a development. The shift from mixed cropping to monoculture has not only affected employment but also soil fertility. Therefore, Vandana Shiva argues that what goes by the name of 'development' – all development strategies typically aimed at promoting Gross National Product, market economy, scientific revolution, colonisation and corporate sectored agriculture – is actually a mal-development process, a source of violence to women and nature throughout the world.

Further, Shiva attacks the Western culture that has evaded Indian agrarian culture and indigenous people's knowledge about nature. It plays a vital role in Indian context because 70 per cent of Indians depend upon traditional systems of production – nature oriented – for their survival and sustenance. Majority of people in the rural areas are involved in natural or organic farming, which involves cultivating different crops and trees and the use of oil-cake as the only pesticide. Moreover, a significant number of these people depend upon small-scale cottage industries and silviculture. The local communities have intricate and close association with the biological diversity of their surroundings to derive sustainable livelihoods.

These observations make Vandana Shiva argue that there is only one path for survival and liberation of nature and woman, and that path is the ecological one – of harmony, sustainability and diversity as opposed to domination, exploitation and surplus. She again asserts that woman plays a vital role in environment, both as its saviour and as a sufferer of ecological mal-development. So, her struggle constitutes a non-violent, non-gendered and humanity-inclusive alternative to the dominant science, technology and development paradigms.

Next, Bina Agarwal (1992), an Indian ecofeminist, examines major elements of ecofeminism in the light of gender and environment. Her perspective differs from the other ecofeminists' perspectives. She raises certain valid questions regarding some of the clearly articulated ecofeminists' arguments such as the connections between the domination of women and exploitation of nature, closeness of women to nature and men to culture and women as vanguards. However, she thinks that women cannot be considered as a homogeneous group. Women differ on the basis of class, race, culture and caste. She argues that gender domination cannot be isolated as the sole cause of degradation of women and nature. She asks, "How can we ignore and be a mute spectator to the political, social and economic domination?" Further, she says that ideological shifts and the disparities that are prevalent in the society have aggravated existing problems.

Agarwal attempts a critical analysis of Vandana Shiva's perspective of ecofeminism. Agarwal contends Shiva's proposition on man-and-nature relationship, which implies that man is separate from nature, and he dominates nature, which is inert and passive. Attention is also drawn towards Shiva's next example of rural women of North-West of India. Shiva has generalised all Third World countries' women under a single category instead of understanding women's heterogeneous identity on the basis of caste, class, race and culture. Bina professes that Shiva, instead of indicating the concrete process of ideological construction of gender and nature and recognising several ideological strands, focuses on the feminine principle as the guiding idea based on the Hindu discourse. The Hindu discourse – Hinduism itself is pluralistic – has not given any clear historical analysis of the practice of the feminine principle and its effect on people and nature.

Then, Shiva ascribes that colonialism and the Western model of development are the root causes of oppression of women and destruction of nature in the Third World countries. She ignores power, privilege and property as well as caste/class stratification and predates colonialism which is the main cause for inequalities and destruction of nature in Third World countries.

By studying various ecofeminists' arguments comprehensively, Bina thinks that ecofeminism is inadequate because as an ideological construction, it establishes the relationships between gender and nature and it becomes only a part of this structure that cannot be considered as a whole structure. This limitation enables Agarwal to come up with an alternative theoretical formulation of ecofeminism, which she terms "Feminist Environmentalism." According to her, it highlights the man-woman relationship with nature, which is rooted in material reality. Gender and class decide division of labour, distribution of property and power, and they structure people's interaction with nature. It shows the effects of environmental changes on people and their response to it. She also focuses on how the division of labour, property and power shape the experience of the people especially the poor peasants, tribal people and cultivators. Along with this, their practical interaction with nature gives them knowledge of nature. Moreover, they are the worst sufferer of the destruction of nature.

Feminist Environmentalism has given a democratic alternative approach that focuses on development and redistribution rather than on welfare. It underlines the need for shifts in the composition of production, alternative technology usage for production, usage of knowledge systems for choices and the class and gender distribution on the basis of production and task, decentralised planning in order to involve localised communities and rural people (especially women) in the interaction with trained scientists.

Another Indian ecofeminist is Chayya Datar, a leading feminist activist and a founding member of 'Forum Against Rape' and '*Stree Mukti Sanghtana.*' In her work *Ecofeminism Revisited* (2011), she puts forth an alternative development model. This development model is opposed to market-oriented capitalism and focuses on decentralisation of power, local production system and subsistence practices especially addressing the problems of rural women whose livelihood depends upon natural resources. Further, she wants to create an awareness of environmental destruction with a special emphasis on a substantial alternative development agenda.

Aruna Gnanadason, formerly at world council of churches, in her article "Tradition of Prudence Lost: A Tragic World of Broken Relationship" (2003), focuses on the loss of age-old agrarian traditions due to globalisation. Her basic contention is against the liberation of economy and the privatisation of every sector, which is replacing the livelihood of villagers. She vehemently says, "What of the life and livelihood of the farming community in India? How is this community going to survive when its intricate web of relationship with the earth is broken?" Further, she posits nuclear reactors and industrial projects impact on the farming community. She contends that the concept of development in the form of dam construction is against the natural phenomenon (Gnanadason 2003:75).

Maithreyi Krishnaraj, a Director of Research Centre for Women's Studies in SNDT Women's University, in *Women Farmers of India* (2008) documents women's close association and dependency on nature. Further, in her anthology *Motherhood in India: Glorification without Empowerment* (2010) explicates sex-gender system, reproduction biology and ideological politics of reproduction capacity of women. The critical overview of Indian ecofeminists accentuates that the impact of ecological degradation is on farming community, due to modern development, in the form of reductionist agriculture, mono cropping, corporate sectored agriculture, etc.

1.4 The genealogy of the Indian novel

Another domain of investigation that the present study deals with is the Indian novel. The novel, relatively a young genre of literary writing, has interesting connections with the making of the Third World culture and politics. Several theoretical works of Homi K. Bhabha (1990), Benedict Anderson (1991), Fredric Jameson (1986) and others have tried to understand the connection between fictional narratives and the nation formation. Besides, we have made several attempts to examine the arrival of the novel from Europe to countries like India. The story of the novel in India is quite interesting and the Indian novel has been studied from different perspectives: generic, thematic, author-oriented studies and the analysis of fictional techniques, etc.

The genesis of the Indian novel can be traced back to nineteenth century British literature. However, the Indian novel has carved a niche for itself in the world literature. The canonical development of Indian novels has a history of more than 150 years. In the middle of the nineteenth century, many Indian litterateurs started to pen down novels. The seeds of novel writing germinated in the literary culture of West Bengal. The advent of novel writing, basically in the regional languages and later in English, was led by Bankim Chandra Chattopadhyay's *Rajmohan's Wife* (1864), and followed by *Durgeshnandini* (1865) and *Kapalkundala* (1866), Rajalakshami Devi's *The Hindu Wife* (1876), Toru Dutt's *Binaca* and so on. As Meenakshi Mukherjee observes, "In most Indian languages – especially in Bengali, Hindi, Marathi, Kannada and Malayalam – the developments occurred in this order, although not simultaneously" (Mukherjee

2016:30). However, by the nineteenth century the novel writing in India was fully evolved. This was made possible by the conditions created by the colonial modernity. To quote Meenakshi Mukherjee again, "By the end of century the novel became the most popular form of print-medium entertainment in at least eight major languages of the country" (Mukherjee 2002:8). The dawn of the nineteenth century witnessed the Nobel laureate Rabindranath Tagore's novel *Choker Bali* (2005), which portrays the plight of a widow in those days (Iyengar 1985).

Later, the novel writing established a strong foundation in the literary culture of India with four major themes – historical, social, psychological and political, among others. The journey of the novel writing began with historical narration. The early historical narratives were historical romances such as Bhudebchandra Mukherjee's *Anguriya Binimoy* (1857) and Bankimchandra Chatterjee's *Singha* (1881) in Bengali; Kishorilal Goswami's *Labangalata* (1891) and Devaki Nandan Khatri's *Chandrakanta* (2017) in Hindi. Simultaneously, in Marathi Hari Narayan Apte produced *Maisorcha Wagh* (1890) and *Gadh Ala Pan Sinha Gela* (1906) and in Kannada Galagnath wrote historical narrative such as *Kumudini* (Mukherjee 2016). Many Indian writers tried to write historical fiction in English: Romesh Chander Dutt's *The Slave Girl of Agra* (1909), Sir Jogendra Singh's *Nur Jahan* (1909), Vimala Raina's *Ambapali* (1962). Manohar Malgonkar's main preoccupation has been historical episodes (Naik 1982, 1985; Iyengar 1962). His major works, such as *Distant Drum* (1960) and *A Bend in the Ganges* (1964), which deal with Indian historical experiences, employ the historical imagination powerfully.

In Bengali, Rabindranath Tagore's *Ghare Baire* (1916) and *Char Adhyay* (1934), Saratchandra Chatterjee's *Patter Dabi* (1926), *Palli Samaj* (1916), and *Arakshaniya* (1916) and in Hindi Premchand's *Sevasadan* (1916) and Bhagavati Charan Verma's *Bhule Bisre Chitra* (1959), in Marathi V. S. Khandekar's *Kanchan Mruga* (1931), *Hirwa Chaha* (1937), and so on are some examples social concerns. Moreover, in English Romesh Chander Dutt's *The Lake of Palms* (1909), Mulk Raj Anand's *Untouchable* (1935), *Coolie* (1936), R. K. Narayan's *The Vendor of Sweets* (1967), Ahmed Ali's *Twilight in Delhi* (1940), and several others deal with social problems of their times.

In addition, a few writers have endeavoured to portray the inner turmoil of individuals in their fiction; those novels could be considered psychological novels. We can cite as examples in Hindi Ilachandra Joshi's *Jahaz Ka Panchi* (1955), Naresh Mehta's *Dubte Mastool* (1954) and in Marathi B S Mardhekar's *Ratricha Divas* (1942) and Vasant Kanetkar's *Ghar* (1951) and also the novels of Anita Desai in English.

Politics forms the crux of Indian novels. The most characteristic feature of the Indian English novels is its engagement with Indian political history. From the early novels such as Tagore's *The Home and the World* (1916) and *Four Chapters* (1934), K. S. Venkataramani's *Kandan, the Patriot* (1932), Raja Rao's *Kanthapura* (1938), Mulk Raj Anand's *The Sword and the Sickle* (1942), K. A. Abbas's *Inqilab* (1975), etc., to the contemporary period such as the works of Shashi Tharoor and Salman Rushdie mainly narrate the Indian political ethos. The barbaric act of partition finds its place in the fictional narratives such as Khushwant Singh's *Train to Pakistan* (1956), Attia Hosain's *Sunlight on a Broken Column* (1961), Chaman Nahal's *Azadi* (1975), Shiv K. Kumar's *A River with Three Banks* (1998), and Manohar Malgonkar's *A Bend in the Ganges* (1960).

The contribution of R. K. Narayan, Raja Rao and Mulk Raj Anand to the making of the novel in India is quite praise worthy. So much has been written on their works and contribution to the Indian English literature that these names have become common among educated Indians. What is unique about the three novelists is that they not only fine-tuned the narrative structure of Indian novels but also participated in the formation of the English language for the novelistic representation of Indian experiences.

Narayan's fictional world consists of more than twelve novels and numerous short stories. His major novels are *Swami and Friends* (1935), *Bachelor of Arts* (1936), *The Dark Room* (1938), *The English Teacher* (1945), *Waiting for Mahatma* (1955), *The Guide* (1958), among others. R. K. Narayan's novels focus on two important forms "(a) recovery of self from delusive experience and (b) self-transcendence, often mediated by agent" (Amur 1985:49). Another prolific writer is Mulk Raj Anand. His major focus is on social, political and cultural issues (Raghavacharyulu 1985) with an emphasis on caste, poverty and backwardness (Iyengar 1962). His novels, for example, *Untouchable* (1935), *Coolie* (1936), *Two Leaves and a Bud* (1937), *The Village* (1939) and *Across the Black Waters* (1940), deal with the problems of Indian social system. Next, the major Indian novelist is Raja Rao, who articulated Gandhi's philosophy "reality into a fictive world, and achieving this end within the framework of an aesthetic form" (Shahane 60), through *Kanthapura* (1938). This novel has been a landmark in the history of Indian writing in English purely from the point of view of its experiment with the English language. Unfortunately, such an attempt was neither continued by Rao himself nor the succeeding novelists. His other novels are *The Serpent and the Rope* (1960), *The Cat and Shakespeare* (1965) and *The Cow of the Barricades* (1947).

Further, Bhabhani Bhattacharaya's *So Many Hungers* (1947), *Music for Mohini* (1952), *He Who Rides a Tiger* (1954), *A Goddess Named Gold* (1960) and *Shadow from Ladakh* (1966) delineate social, economic and political changes in India through the portrayal of poverty, hunger and exploitation (Iyengar 1985). Further, he also espouses "the freedom struggle, the fight against poverty, superstition, caste-domination, modernization through industrialization, East-West encounter, the struggle to establish a new democratic and socialistic order" (Desai 1985:119).

The Indian English novel's journey marched ahead with the post-colonial and diasporic novelists like Amitav Gosh, Shashi Tharoor, Vikram Seth, Ruskin Bond, Shashi Despande, Kamala Markandaya and Arundhati Roy along with diasporic novelists such as Kiran Desai, Jhumpa Lahiri, Agha Shahid Ali, Rohinton Mistry, Salman Rushdie and so on. The legacy of the Indian novel writing is carried forth by the corporate world novelists such as Chetan Bhagat, Amish Tripathi and others.

The above-sketched narrative of the Indian novel in general and the Indian-English novel in particular illustrates the rise and development of the novel form in India, its major thematic preoccupations. However, Indian novels have been studied from variety of perspectives. Especially, they have been largely studied from social (Bhatnagar 2001; Mukherjee 2002), political (Kaushik 1988, 2001; Prasad 2001; Bhatnagar 2007), historical (Naik 1982, 1985; Iyengar 1962) and cultural perspectives (Ganguli 1977; Lannoy 1971). In addition, regional novels have been studied and researched by the critics of their respective languages. It is difficult to render here a review of these studies, as most of them are not available in English.

1.5 The discursive formation of the environment in Indian novel

Along with the generic evolutionary development of the Indian novel, it is quite evident that creative writers have also considered ecological problems and have given vent to them in their literary renderings. The critical discourse on ecofeminism makes a novel attempt to understand the relationship between literary texts and problems with regard to gender and ecological crisis. It is also necessary to study the discursive formation of the environment in the Indian literary narratives.

1.5.1 Ancient literary narratives

As far as the Indian context is concerned for literary narratives, the relationship between nature and literature could be traced back to our epic tradition and Indian scriptures such as *Rgveda*, *Upanishada*, *Puranas* and epics along with the classical Indian texts such as Kalidasa's *Ritu Sanhar*, *Meghdoot* and *Abigyaan Sakuntalam* project the generative connections between man and nature. These ancient literary narratives have projected inextricable and intimate bonding of man and nature. In the *Rgveda*, the 90th *sukta* of the first mandala subsumes that the human beings consider nature as their deity and revere it by seeking its blessings for the smooth execution of their lives:

> madhu vātā ṛtāyate madhu kṣaranti sindhavaḥ |
> mādhvīrnaḥ santvoṣadhīḥ | |
> madhu naktamutoṣaso madhumat pārthivaṃ rajaḥ |
> madhu dyaurastu naḥ pitā | |
> madhumān no vanaspatirmadhumānastu sūryaḥ |
> mādhvīrghāvo bhavantu naḥ | |
> śaṃ no mitraḥ śaṃ varuṇaḥ śaṃ no bhavatvaryamā |
> śaṃ na indro bṛhaspatiḥ śaṃ no viṣṇururukramaḥ | |

The *Sukta* explicates that the human beings pray all natural objects, wind, river, sun, plants and cows, to bless them with prosperity by being sweet and kind. Like this, the *Vedas* provide many evidences to culminate that how nature was constant companion of human beings and their homage to nature as a deity. In the *Atharvaved*, the most loaded hymn *Bhumisukta*, vivid descriptions of the earth and its gift to humankind adds nuances to the above perspective:

> *yasyaam Samudra Uta Sindhur-Aapo Yasyaam-Annam Krssttayah Sambabhuuvuh | Yasyaam-Idam Jinvati Praannad-Ejat-Saa No Bhuumih Puurva-Peye Dadhaatu.*

It describes that the mother earth is always bestowed on us with water and food by which all living beings are alive. She should be kind and generous to gift the same thing in the future. All Vedas are especially valorises the basic elements of environment: *Apah* (water), *Vayu* (air), *Akasha* (sky), etc. These scriptures further propound that human beings and nature are inseparable and integral part of each other and God dwells in every particle of the nature. Sri Krishna says in the *Bhagavad-Gita*, "My energy enters the earth, sustaining all that lives: I become the moon, giver of water and sap, to feed the plants and the trees." The ensemble of these scriptures represents synergistic symphony of human beings and nature.

The *Taittiriya Upanishad* says, "Creating all things, Brahman entered into everything. Entering into all things, he became that which has shape and that which is shapeless; he became that which can be defined and that which cannot be defined." Analogous to this, the *Brhadaranyaka Upanishad* says, "This earth is like honey for all beings, and all beings are like honey for this earth. The intelligent, immortal being, the soul of this earth, and the intelligent immortal being, the soul in the individual being – each is honey to the other. Brahman is the soul in each; he is indeed the Self in all. He is all." These texts manifest that all the creatures on the earth have harmonious and reciprocal relationship. In the *Manusmriti*, Manu bespeaks that trees and plants were full of consciousness and they experienced pleasure and pain (Vrajaprana 2013).

Besides, the ancient literature recounts that the idea of a developed society germinated in the forest (Tagore 2018). The significance of forest is very vividly explicated in the Vedas. In *Rigveda*, *Aranyanisukta* and *Oshadhi Sukta* address forest as a mother, "O Mother! Hundreds are your birth places and thousands are your shoots." The *Chandogya Upanishad* espouses, "Water has generated plants which in turn generated food." The *Atharvaveda* elaborates, "the earth is keeper of creation, container of forests, trees and herbs.'" In the same line *Purana* says, "one tree is equal to ten sons" (Anilkumar 96). Such ecological concerns can be traced throughout Indian literary history.

Further, Indian great epics, the *Ramayana* and the *Mahabharata*, embed the forest life with human life as a major narrative structure. The *Aranya Kaanda* (The Section on Forest) and the *Vanavaasa* (a life in forest) are classic illustrations of writings on nature. Veda Vyasa, in *Mahabharata*, had elaborately narrated the episode *Vanaparvan* (the period in the forest) that is the Pandavas' 12-year stay in the forest. This narration not only describes a particular episode but also gives the evidence to the fact that is the biological diversity and interdependence and interrelationship of human beings and nature. Besides, these texts are prominent landscape for the researcher to examine various environmental concepts such as environmental management, natural justice and environmental ethics.

Later, one can trace the holistic relationship of human beings and nature in classical literary narrations. In the fourth century A.D., the renowned Indian classic Sanskrit dramatist Kalidas wrote *Shakuntala*, a great play not only in Sanskrit literature, but also in the literature of the world. The theme of the drama is borrowed from the *Mahabharata*. The play revolves around two characters: Dushyanta, a celebrated king, and Shakuntala, the daughter of Menka (the celestial nymph). She was abandoned by her parents and reared by the sage Kanva. At the opening of the play, Dushyanta, on a hunting expedition, arrives at the Kanva hermitage. He marries Shakuntala. After a few days, when Shakuntala leaves the *Ashram* to meet Dushyanta, the entire *Ashram* plunges into sorrow. Kanva, Anusuya and all shed tears at the departure of Shakuntala, even the trees, plants and birds bow down with grief. The entire play is epitome of the harmonic relationship of human beings and nature. In this context, it is very apt to contemplate on Swami Vivekananda's perspectives on true universality:

> When I say I am separate from you it is a lie, a terrible lie. I am one with the universe. . . . I am one with the air that surrounds me, one with heat, one with light, eternally one with the whole Universal Being, who is called this universe, who is mistaken for this universe, for it is He and nothing else. . . . I am one with That.
>
> (Paranjape 2015)

1.5.2 The contemporary environmental literary narratives

Mapping the contour of the ancient environmental narratives in the Indian literary canon, it is interesting to observe that several contemporary Indian novels are sensitive to the ecocritical and ecofeminist issues. To mention a few Munshi Premchand's *Godan* (1936), Jim Corbett's *Man-Eater of Kumaon* (1989), Kamala Markandaya's *Nectar in a Sieve* (2009), Ruskin Bond's *The Blue Umbrella* (2013) and *The Cherry Tree* (1996), O V Vijayan's *Madhuram Gayathi* (1990), Rohinton Mistry's *Such a Long Journey* (1991), Gita Mehta's *A River Sutra* (1994) and Arundhati Roy's *The God of Small Things* (2012). Roy's five volumes of non-fiction writings, *The Algebra of Infinite Justice* (2001), *An Ordinary Person's Guide of Empire* (2005), *Listening to Grasshoppers* (2009) and *Broken Republic* (2011); Sohaila Abdulali's *The Madwoman of Jogare* (1998); Anita Nair's *Better Man* (1999); Amitav Ghosh's *The Hungary Tide* (2004), *Sea of Poppies* (2008) and *River of Smoke* (2008); Anuradha Roy's *An Atlas of Impossible Longing* (2008); Mamang Dai's *The Legends of Pensam* (2006),; Tamsula Ao's *These Hills Called Home Stories from a War Zone* (2006) and *Laburnum for My Head Stories*

(2009); Usha K.R.'s *Monkey-Man* (2010); Ruchir Joshi's *The Last Jet-Engine Laugh* (2012); Purnachandra Tejasvi's *Carvalho* (2016) have explicated the web of relationship among human beings, wildlife and nature as well as the impact of human development attitude on nature.

Apart from the novels, the other literary genres exhibit the aspects of environment in their narration. The major nature-loving poets are from the Northeast region of India. Mamang Dai's 'River' and 'The Missing Links,' Dayananda Pathak's 'Coral Island,' Tamsula Ao's 'The Nightingale of North-eastern India' and so on are some of the examples. Along with these, a few mainstream poets have tried to show natural phenomenon and ecological erosion. For example, Toru Dutt's 'The Lotus' and 'Our Casuarina Tree,' S K Chettur's 'Red Lotus,' Rabindranath Tagore's 'Breeezy April,' Sarojini Naidu's 'Summer Woods,' Dilip Chitre's 'The Felling of the Banyan Tree,' Gieve Patel's 'On Killing a Tree' and Keki Daruwalla's 'Boat-Ride Along the Ganga.' The brief catalogue of the narration of environment demonstrates the discursive formation of environment in Indian literary genre. The critical examination of all these texts place on the record that the ancient scriptures describes innumerable natural objects and their relationship with the human beings, which is the legacy of their healthy relationship – nowhere these texts document the deterioration and degradation of nature. On the other hand, the contemporary literary narratives are the archive of the loss of flora and fauna and the loss of reciprocal healthy relationship of human beings and nature because of human development attitude. Though there is seismic shift in the narration, ultimately nature is a backdrop for all literary narration. This brief spectrum of environmental narration is interesting to note that literary texts of both classical and contemporary epochs have taken the life of nature as their main concern.

Although the ecological crisis and feminism form the backdrop of several Indian novels, these have hardly been studied from the perspective of ecofeminism. Therefore, this book is an exploration of ecofeminist discourse through the select Indian novels. The novels such as Kamala Markandaya's *Nectar in a Sieve* (2009)and *The Coffer Dams* (2008), Shivram Karanth's *Return to Earth* (2002), Pundalik Naik's *The Upheaval* (2002), Indira Goswami's *The Man from Chinnamasta* (2006), Indra Sinha's *Animal's People* (2007), Sarah Joseph's *Gift in Green* (2011) and Na D'Souza's *Dweepa* (2013) are mainly concerned with the problem of ecological crisis and its impact on human beings. No attempt has been made so far to group together these Indian novels as narratives of ecocritical and feminist narratives when such a reading gives a different identity to the Indian novel.

1.6 Rumination on Indian environmental movements and protests

After a comprehensive review of the discursive formation of environment in literary narratives, for proper orientation of this underlying assumption in

India, we need to examine synchronic snapshots of Indian environmental crisis by putting forth a question to ruminate: are Indian environmental problems gender related?

India also witnessed a large number of environmental movements and protests. In India, environmental problems are not solved in isolation – gender analysis – because natural resources are the effective means of survival for the majority of Indians. The first epoch-making Chipko movement led the peasants of the Garhwal Himalaya region to protest against the commercial felling of trees by hugging the trees (Gadgil & Guha 1994; Rao 2012). Himalaya is dying due to the onslaughts of aggressive developments in the form of damming the rivers, deforestation, mining and luxury tourism. The basic reason for the Chipko movement was minerals, soils and forest of the Uttarakhand region, which attracted many entrepreneurs. This movement had taken birth in 1964 in Gopeshwar in Chamoli district in the form of *Dasoli Gram Swarajya Sangh* (DGSS). This organisation was founded by Chandi Prasad Bhatt, an Indian Gandhian environmentalist and social activist who was one of the pioneer protesters. The initiation of the first protest of Chipko movement was against the local operators who wanted to cut the Ash trees. Later, the Ash trees were sold to a sports goods manufacturing company for the purpose of making bats and tennis rackets. The villagers initially appealed government to stop the deforestation; but their plea fell on the deaf ears. Therefore, the villagers adopted a non-violent protest against the felling of the trees. Indeed, the Chipko movement was led by women hugging the trees to save them. However, men also play very vital and prominent role in this protest including Gaura Devi, Bachni Devi, Chandi Prasad Bhatt, Sunderlal Bahuguna, Govind Singh Rawat, Doom Singh Negi, Ghanasyam Raturi, etc. In historical milieu of the movement, Bahuguna says,

> It was in 1930 that people in our area revolted against the commercialisation of the forests. To suppress that rebellion, on 30th May 1930 the army was sent by the rulers of the State. As many as 17 persons were shot dead, about 80 arrested. Though the movement was then suppressed, we got inspiration from them. We established a memorial to those martyrs. In 1969, we repeated a pledge in front of their memorial. This became the background of Chipko.
>
> (Bahuguna 1983:17)

Though the Chipko Movement was led by women to save the forest from the government's project of mining, men also participated in this movement. The prominent among them was Sunderlal Bahuguna. As he says,

> The objective of this policy should be to heal the wounds of the Himalaya, keep it as a place to live for the local inhabitants and accessible to nature lovers and spiritual seekers, use natural resources in a sustainable manner to achieve regional self-sufficiency, keep the landscape intact, protect

> biodiversity and establish local autonomy for the advancement of culture. This will save both the nature and culture of these great mountains, a source of varied inspiration to humankind.
>
> (Bahuguana 1983:18)

Therefore, the protest was not gender oriented, it was human beings oriented. There were foot marches of 4870 km from Kashmir to Kohima. It took 300 days to create the awareness of the impact of deforestation on their livelihood. In addition to this, the Appiko movement, the Western Ghats of Karnataka, started by Pandurang Hegde to save the trees, form monoculture farming (Gadgil & Guha 1994).

Another grassroots movement that launched a protest against the building of a dam on the Narmada River – which is considered to be one of the most catastrophic environmental disasters in the world – was the *Narmada Bachao Andolan* (Save the Narmada River Movement) (Rao 2012). In this protest men and women equally participated, the crusader is Medha Patkar, as Bahuguna says, "When the work on the dam started in 1978, many men and women went to stop the work and were arrested and sent to jail. The whole area was converted into a police cantonment so that the people could not do anything. The government said that they would hold talks, but nothing came out of them." Further, Bahuguna adds, "We have been camping in a hut for last four years near the dam site in non-violent protest and have been able to stop the work twice. Twice I fasted to make the government realize the need for a review of the technical, social, economic, cultural, ecological and spiritual aspects of the project." Next, there emerged various protests against mining projects, for example, in 1947 in the Doon valley in northwest India, and more recently in 1983 the Gandhamardan hills of Sambalpur district in Odisha. Such struggles sought to draw our attention to the irreversible consequences such as deforestation, the drying up of water resources, and loss of agrarian lands (Gadgil & Guha 1994). Along with this, Tehri in the east, Tungabhadra in the south and the Koyana project in Maharashtra raised their voices to stop the construction of dams (Datar 2011).

A few protests for displacement are *Jan Andolan* against Dabhol Power Plant in Maharashtra, the anti-nuclear protest against Kaiga project in Karnataka, anti-mine protest in Dehradhun valley and protests against missile test in Billiapal (Orissa) and Nethral (Bihar) and against tourism are *Himalaya Bachao Andolan* in the north and Bailancho Saad in Goa (Datar 2011).

In addition to these, Save Silent Valley, a social movement, was started in 1973 to save an evergreen tropical forest in the Palakkad district of Kerala from a hydroelectric project (Chengappa 2009). The landfilling of hundreds of acres of water-beds and mangrove forests in Ernakulam district was to build a cricket stadium which created a lot of pernicious problems (Joseph 2011). Besides, in the year 1984 the pesticide factory of Bhopal – owned by the multinational company Union Carbide – accidentally released deadly toxic gas, which had taken a heavy toll on the lives of many people and caused massive damage to plant and animal lives. And there were long-lasting aftermath effects on survivors (Mukherjee 2010).

Moreover, in the year 2007, a malicious incident took place, when the state government of West Bengal signed a memorandum of understanding with the Salim industrial group of Indonesia to build a gigantic 'chemical hub' on the cultivated land. These environmental movements and protests characterises that the ecological crisis are because of the modern developmental attitude and to resolve these problems men and women both joined their hands together.

1.7 Conclusion: the formulation of framework

The critical investigations of the above-mentioned ecofeminist perspectives illuminate the problems associated with ecology and women. As evident in the above narrative, the ecofeminist discourse is praxis-oriented and heterogeneous; however, it engages mainly with three basic elements such as patriarchy, feminism and environmental problems. This theory affirms that modernisation, globalisation (Eaton & Lorentzen 2003), developments in science and technology (Merchant 1980; Griffin 1980; Mies & Shiva 1993), reproduction technology (Adams 1994, 2010) and corporate agriculture (D'Eaubonne 1974; Shiva 2010) are the basic reasons for environmental crisis. Further, as per most renowned ecofeminists, man, understood as a patriarchal power, operates through these processes in ruining the ecological and feminine life (D'Eaubonne 1974 1974; Daly 1978; Merchant 1980; Griffin 1980; Warren 1997; Plumwood 1993; Adams 2010; Mies & Shiva 1993).

The discussion of many ecofeminists put forth three basic tenets of ecofeminism: (1) there is a connection between women and nature based on the ideas that (a) women have close association with nature because of their dependency on it; and (b) the development of science and technology is the cause of exploitation of women as well as they are in the vanguard to protect, preserve and nurture nature (D'Eaubonne 1974; Ruether 1975; Daly 1978; Merchant 1980; Mies & Shiva 1993); (2) women are life givers, saviours and nurturers of nature (D'Eaubonne 1974; Ruether 1975; Daly 1978; Merchant 1980; Mies & Shiva 1993); and (3) patriarchy in the name of development is the root cause of the exploitation of women and degradation of nature (D'Eaubonne 1974; Ruether 1975; Daly 1978; Warren 1997; Salleh 1993; Griffin 1980; Mies & Shiva 1993; Mies 1986; Birkeland 1993). In the light of this understanding, the forthcoming chapters analyse Indian novels to comprehend ecofeminist discourse in Indian context.

Bibliography

Abrams, M.H. *A Handbook of Literary Terms*. Cengage Learning India, 2009.

Adams, Carol. *Neither Man Nor Beast: Feminism and the Defense of Animals*. Continuum, 1994.

———. *The Sexual Politics of Meat*. Continuum, 2010.

Agarwal, Bina. "Environmental Management, Equity and Ecofeminism: Debating India's Experience." *Journal of Peasant Studies*, vol.25, no.4, 1998, pp. 55–95.

———. "The Gender and Environment Debate: Lessons from India." *Feminist Studies*, vol.18, no.1, 1992, pp. 119–158.

Allister, Mark Christopher. *Refiguring the Map of Sorrow*. University Press of Virginia, 2001.

Amur, G.S. "A Saint for Malgudi: A New Look at R. K. Narayan's the Guide." *Perspectives on Indian Fiction in English*, edited by M.K. Naik. Abhinav Publications, 1985.

Anderson, Benedict. *Imagined Communities: Reflections on the Origin and Spread of Nationalism*. Verso, 1991.

Andrea, Campbell. "The Gender and Environment Debate: Lessons from India." *Feminist Studies*, vol.18, no.1, 1992, pp. 119–158.

———. *New Direction in Ecofeminist Literary Criticism*. Cambridge Scholars Press, 2008.

Anilkumar, A.R. *Culture and Environmental Science in Sanskrit*. BA Sanskrit IV Semester Core Course, University of Calicut, India.

Bahuguna, Sundarlal. *Dharati Ki Pukar*. Radhakrishna Prakashan, 2007.

———. *Walking with the Chipko Message* (Teheri Garhwal District). Navijivan Asharam, 1983.

Barry, Peter. *Beginning Theory: An Introduction to Literary and Cultural Theory*. Manchester University Press, 2002.

Basham, A. L., editor. *A Cultural History of India*. Oxford University Press, 1975.

Bate, Jonathan. *The Green Studies Reader: From Romanticism to Ecocriticism*. Routledge, 2000.

———. *A Song of the Earth*. Harvard University Press, 2000.

Baviskar, Amita. *The Belly of the River: Tribal Conflicts over Development in the Narmada Valley*. Oxford University Press, 1995.

Bhabha, K. Homi. *Nation and Narration*. Routledge, 1990.

Bhasin, Kamla. *Understanding Gender*. Kali for Women, 2000.

Bhatnagar, Manmohan. *Indian Writing in English Volume II*. Atlantic Publishers, 2001.

Bhatnagar, O.P. *Indian Political Novel in English*. Sarup & Sons, 2007.

Biehl, Janet. *Rethinking Ecofeminist Politics*. South End Books, 1991.

Birkeland, Janis. "Ecofeminism: Linking Theory and Practice." *Ecofeminism: Women, Animals, Nature*, edited by Greta Gaard. Temple University Press, 1993.

Bookchin, Murray. *Remaking Society*. Black Rose, 1989.

Buell, Lawrence. *The Environmental Imagination: Thoreau, Nature Writing and the Formation of American Culture*. Princeton University Press, 1995.

———. *Writing for an Endangered World Literature, Culture, and Environment in the U. S. and Beyond*. Harvard University Press, 2001.

Carlassare, Elizabeth. "Essentialism in Ecofeminist Discourse." *Ecology*, edited by Merchant Carolyn. Humanities Press, 1994.

———. "Socialist and Cultural Ecofeminism: Allies in Resistance." *Ethics and the Environment*, vol.5, no.1, 2000, pp. 89–106.

Carson, Rachel. *The Silent Spring*. Penguin Books, 1963.

Chengappa, Raj. "1976: Silent Valley Movement: The Genesis of Green." *India Today*, 24 December, 2009.

Daly, Mary. *Gyn/Ecology: The Metaethics of Radical Feminism*. Beacon Press, 1978.

Datar, Chhaya. *Ecofeminism Revisited*. Rawat Publications, 2011.

D'Eaubonne, Francoise. "Feminism-Ecology: Revolution or Mutation?" *Ethics and the Environment*, vol.4, no.2, 1999a, pp. 175–177.

———. *Le feminism ou la mort*. Pierre Horay, 1974.

———. "What Could an Ecofeminist Society Be?" *Ethics and the Environment*, vol.4, no.2, 1999b, pp. 179–184.

Desai, S.K. "Bhabhani Bhattacharya: The Writer Who Rides a Tiger." *Perspectives on Indian Fiction in English*, edited by M.K. Naik. Abhinav Publications, 1985.

D'Souza, Na. *Dweepa*. Translated by Susheela Punitha. Oxford University Press, 2013.

Eaton, Heather, and Lois Ann Lorentzen, editors. *Ecofeminism and Globalisation*. Rowman & Littlefield Press, 2003.

Fukuoka, Masanobu. *The One-Straw Revolution: An Introduction to Natural Farming*. NYRB Classics, 1975.

Gaard, Greta. *Ecofeminism and Native American Cultures in Ecofeminism*. Temple University Press, 1993b.

———. "Ecofeminism Revisited: Rejecting Essentialism and Re-Placing Species in a Material Feminist Environmentalism." *Feminist Formations*, vol.23, no.2, 2011, pp. 26–53.

———, editor. *Ecofeminism: Women, Animals, Nature*. Temple University Press, 1993a.

Gaard, Greta, and P.D. Murphy, editors. *Ecofeminist Literary Criticism: Theory, Interpretation, Pedagogy*. University of Illinois Press, 1998.

Gadgil, Madhav, and Ramchandra Guha. "Eçological Conflicts and the Environmental Movement in India." *Development of Change*, vol.25, 1994, pp. 101–136.

Ganguli, B.N., editor. *Tradition, Modernity and Development: A Study in Contemporary Indian Society*. Macmillan, 1977.

Garrard, Greg. *Ecocriticism*. Routledge, 2004.

Glotfelty, Cheryll, and Harold Fromm. *The Ecocriticism Reader Landmarks in Literary Ecology*. University of Georgia Press, 1996.

Gnanadason, Aruna. "Traditions of Prudence Lost: A Tragic World of Broken Relationships." *Ecofeminism and Globalization*, edited by Heather Eaton and Lois Ann Lorentzen. Rowman & Littlefield Press, 2003, pp. 73–87.

Goldman, Michael, and Rachel Schurman. "Closing the 'Great Divide': New Social Theory on Society and Nature." *Annual Review of Sociology*, vol.26, 2000, pp. 563–584.

Goswami, Indira. *The Man from Chinnamasta*. Translated by Prashant Goswami. Katha, 2006.

Griffin, Susan. *Women and Nature: The Roaring Inside Her*. Harper & Row Publishers, 1980.

Haeckel, Ernst. *Generel lemorphologie der organismen* (General Morphology of the Organisms). G. Reimer, 1866.

Hobgood-Oster, Laura. "Ecofeminism: Historic and International Evolution." *Encyclopedia of Religion and Nature*, edited by Bron Taylor. Continuum, 2005.

Iyengar, Srinivasa. *Indian Writing in English*. Asia Publishing House, 1962.

———. "On Re-Reading the Serpent and the Rope." *Perspectives on Indian Fiction in English*, edited by M.K. Naik. Abhinav Publications, 1985.

Jameson, Fredric. "Third-World Literature in the Era of Multinational Capitalism." *Social Text*, vol.15, 1986, pp. 65–88.

Joseph, Sarah. *Gift in Green*. Translated by Valson Thampu. Harper Perennial, 2011.

Karanth, Shivram. *Marali Mannige*. Sapna Book House, 2015.

———. *Return to Earth*. Translated by Padma Ramachandra Sharma. Sahitya Akademi Centre for Translation, 2002.

Kaushik, Asha. *Political Aesthetics and Culture: A Study of Indo-Anglican Political Novel*. Manohar, 1988.

———. *Politics, Symbols, and Political Theory: Rethinking Gandhi*. Rawat, 2001.

Kerridge, Richard. "Small Rooms and the Ecosystem Environmental and DeLillo's White Noise." *Writing the Environment*, edited by Richard Kerridge and Neil Sammells. Zed Books, 1998.

King, Ynestra. "The Eco-Feminist Perspective." *Reclaiming the Earth: Women Speak Out for Life on Earth*, edited by L. Caldecott and S. Leland. Women's Press, 1983a.

———. "The Ecology of Feminism and the Feminism of Ecology." *Healing the Wounds: The Promise of Ecofeminism*, edited by Judith Plant. New Society Publishers, 1983b.
———. "Engendering a Peaceful Planet: Ecology, Economy, and Ecofeminism in Contemporary Context." *Women's Studies Quarterly: Rethinking Women's Peace Studies*, vol.23, no.3/4, 1995, pp. 15–21.
Krishnaraj, Maithreyi. *Motherhood in India: Glorification without Empowerment*. Routledge, 2010.
Krishnaraj, Maithreyi, and Aruna Kanchi. *Women Farmers of India*. National Book Trust, 2009.
Lannoy, Richard. *The Speaking Tree: A Study of Indian Culture and Society*. Oxford University Press, 1971.
Luniya B. N. *Life and Culture in Ancient India*. Lakshmi Narani Agarwal, 1982.
Markandaya, Kamala. *The Coffer Dams*. Penguin Books, 2008.
———. *Nectar in a Sieve*. Penguin Books, 2009.
———. *Pleasure of City*. Penguin Books, 2011.
Merchant, Carolyn. *The Death of Nature*. HarperCollins Publishers, 1980.
———. *Earthcare: Women and the Environment*. Routledge, 1995.
Mies, Maria. *Patriarchy & Accumulation on a World Scale*. Zed Books Ltd, 1986.
Mies, Maria, and Vandana Shiva. *Ecofeminism*. Rawat Publications, 1993.
Mishra, D.D. *Fundamental Concepts in Environmental Studies*. S. Chand, 2008.
Mishra, R. "Form, Function and Factors in Ecology." *Journal of the Indian Botanical Society*, vol.46, 1967, pp. 144–153.
Mukherjee, Meenakshi. *Early Novels in India*. Sahitya Akademi, 2002.
———. *Realism and Reality: The Novel and Society in India*. Oxford University Press, 1985.
———. *The Twice Born Fiction*. Arnold-Heinemann Publishers, 2016.
Mukherjee, Upamanyu Pablo. *Postcolonial Environments: Nature, Culture and the Contemporary Indian English Novels*. Palgrave Macmillan, 2010.
Naess, Arne. *Ecology, Community and Lifestyle*. Translated and edited by David Rothenberg. Cambridge University Press, 1989.
Naik, M.K. *A History of Indian English Literature*. Sahitya Akademi, 1982.
———, editor. *Perspectives on Indian Fiction in English*. Abhinav Publications, 1985.
Naik, Pundalik. *The Upheaval*. Translated by Vidya Pai. Oxford University Press, 2002.
Nandy, Ashis, editor. *Science, Hegemony and Violence: A Requiem for Modernity*. Oxford University Press, 1988.
Odum, Eugene P. *Ecology: A Bridge between Science and Society*. Sinauer Associates, 1996.
———. *Fundamentals of Ecology*. Saunders, 1953.
Paranjape, Makarand. *Swami Vivekananda: A Contemporary Reader*. Routledge, 2015.
Plato. *The Republic*. Edited by Desmond Lee. Penguin, 1987.
Plumwood, Val. "Ecofeminism: An Overview and Discussion of Positions and Arguments." *Australian Journal of Philosophy*, vol.64, 1986, pp. 120–138.
———. *Environmental Culture: The Ecological Crisis of Reason*. Routledge, 2002.
———. *Feminism and the Mastery of Nature*. Routledge, 1993.
———. "Nature, Self, and Gender: Feminism, Environmental Philosophy, and the Critique of Rationalism." *Journal of Hypatia, Inc and Wiley*, vol.6, no.1, 1991, pp. 3–27.
Prasad, Amarnath. *Critical Response to Indian Fiction in English*. Atlantic Publishers, 2001.
Raghavacharyulu, D.V.K. "Small Scale Reflection on a Great House of Fiction: R. K. Narayan's Chronicles of Malgudi." *Perspectives on Indian Fiction in English*, edited by M.K. Naik. Abhinav Publications, 1985.
Rao, Manisha. "Ecofeminism at the Crossroads in India: A Review." *DEP*, vol.20, 2012, pp. 124–142.

Roszak, Theodore. *The Voice of the Earth*. Simon & Schuster, 1993.
Rueckert, William. "Literature and Ecology: An Experiment in Ecocriticism." *Lowa Review*, vol.9, no.1, 1978, pp. 71–86.
Ruether, Rosemary Radford. "Ecofeminism: First and Third World Women." *American Journal of Theology & Philosophy*, vol.18, no.1, 1997, pp. 33–45.
———. *New Women/New Earth: Sexist Ideologies and Human Liberation*. Seabury, 1975.
Sainath, P. *Everybody Loves a Good Drought*. Penguin Books, 1996.
Salleh, Ariel. "Class, Race and Gender Discourse in Ecofeminism and Deep Ecology Debate." *Environmental Ethics*, vol.15, no.3, 1993, pp. 225–244.
———. "Ecofeminism as Sociology." *Capitalism Nature Socialism*, vol.14, no.1, 2003, pp. 61–74.
Sandilands, Catriona. *The Good Natured Feminist*. University of Minnesota, 1999.
Selvamony, Nirmal. "Considering the Humanities Ecotheoretically." *Journal of Contemporary Thought*, vol.40, 2014, pp. 5–19.
Selvamony, Nirmal, et al. *Essays in Ecocriticism*. Sarup & Sons and OSLE-India, 2008.
Shahane, Vasant. "Fiction and Reality in Raja Rao." *Perspectives on Indian Fiction in English*, edited by M.K. Naik. Abhinav Publications, 1985.
Shiva, Vandana. *Biopiracy*. Natraj Publishers, 2012.
———. *Staying Alive Women, Ecology and Survival in India*. Women Unlimited, 2010.
———. *The Violence of the Green Revolution*. Zed Books Publication, 1993.
Sinha, Indra. *Animal's People*. Simon & Schuster, 2007.Tagore, Ravindranath. *Choker Bali*. Rupa, 2005.
———. *Prachin Sahitya*. CreateSpace Independent Publishing, 2018.
Tejasvi, Purnachandra. *Carvalho*. Pustak Prakashan, 2016.
Vrajaprana, Pravrajika. "Global Ecology and Vedanta: Part 1 & 2." Ramakrishna Mission Institute of Culture, March 2013, posted at Vedanta Society of Southern California.
Warren, J. Karen, editor. *Ecofeminism: Women, Culture, Nature*. Indiana University Press, 1997.
———. "Ecofeminist Philosophy and Deep Ecology." *Philosophical Dialogues*, edited by Nina Witoszek and Andrew Brennan. Rowman & Littlefield, 1999.
———, editor. *Ecofeminist Philosophy: A Western Perspective on What It Is and Why It Matters*. Rowman & Littlefield Press, 2000.
———. *Ecological Feminism*. Routledge, 1994.
———, editor. *Ecological Feminist Philosophies*. Indiana University Press, 1996.
———. "The Power and Promise of Ecological Feminism." *Environmental Ethics*, vol.12, no.3, 1990, pp. 125–146.
Warren, J. Karen, and Dunane L. Cady. "Feminism and Peace: Seeing Connections." *Hypatia*, vol.9, no.2, Spring 1994, pp. 4–20.

2 Narratives of agriculture

Nectar in a Sieve, The Upheaval, Return to Earth and *Gift in Green*

2.1 Introduction

We have discussed elaborately, in the previous chapter, the critical reflections, arguments and perspectives of Indian environmental and ecofeminist discourse. Accordingly, the developmental attitude is responsible for the ecological crisis. The markers of development such as the construction of mega cities and commercial endeavours like tannery, mining, colonial modernisation among others have made a huge impact on the lives of common man and nature. The worst victims of these structures of development are communities in the domains of agriculture and pastoral. Against this backdrop, the present chapter examines Kamala Markandaya's *Nectar in a Sieve* (2009), Pundalik Naik's *The Upheaval* (2002), Shivram Karanth's *Return to Earth* (2002) and Sarah Joseph's *Gift in Green* (2011) as narratives of agriculture. The setting of these novels, by and large, is agrarian and pastoral life. They, with a little bit of variation, show how the idyllic life of agriculture is disturbed by modern developmental attitude. The reading of these novels not only shows agriculture as an ecofeminist concern but also add a new dimension to this discourse.

2.2 *Nectar in a Sieve*: the impact of tannery on pastoral life[1]

Kamala Markandaya is an important post-independent Indian novelist and short-story writer. She had an illustrious career in literary writing, which spanned over four decades, as it started with *Nectar in a Sieve* (2009) and ended with the novel *Pleasure of City* (2011). Like the Western women novelists, Indian women novelists have also established the tradition of feminine sensibility through their narration, which is an important perspective of the modern Indian Renaissance. Though Markandaya highlights variegated themes, these revolve around feminine sensibility. Most of her novels concentrate on human psychology, crisis in human relationships, the conflict between the East and West, tradition and modernity, rural and urban life, etc. Many of her novels portray realism and ruminate on the economic, socio-cultural and political aspects. Stephen Hemenway says, "She is definitely one of the most productive, popular and skilled Indo-Anglian novelists and a superb representative of the growing number of Indian women writing serious literature in English" (Hemenway 1975:52). She has 10 novels and many short stories to her credit. Her first novel *Nectar in a Sieve* (2009) was widely acclaimed. Her works include *Inner Fury* (1955), *A Silence of Desire* (1960), *Possession* (1963), *A Handful of Rice* (1966), *The Coffer Dams* (2008), *The Nowhere Man* (1972), *Two Virgins* (1973), *The Golden Honeycomb* (1977) and *Pleasure City* (1982).

Kamala Markandaya's *Nectar in a Sieve* (2009) is a classic pastoral narrative. It narrates the story of a small South Indian village. The prominent feature of the novel is to represent the peasant lives – their activities, hardships, aspirations and sorrows, and the impact of tannery on their lives. The novel centres on Rukmani and her family. Rukmani, the protagonist of the novel, is the daughter of a former headman of the village; she marries a poor tenant farmer Nathan. The first six years of her married life are spent without much difficulty. Rukmani spends all her savings on the marriage of her daughter Irawaddy. Later, the introduction of tannery and the rise in the price of daily commodities compel Rukmani to sell her cattle. Moreover, Arjun and Thambi, her sons, join tannery; eventually, they lose their jobs in tannery and go to Ceylon in search of work. In order to pay dues to the landowner, Rukmani and Nathan sell household material and bullocks. Their third son, Raja, dies in a combat with the tannery people. Owing to nature's fury, Nathan and Rukmani lose their crops, which leads to hunger and starvation. Kuti, their son, becomes a victim of starvation. Irawaddy becomes a prostitute to save her brother from hunger. The landowner, Sivaji, sells land to the tannery people. Being landless, Rukmani and Nathan go to the city to meet their son. But unfortunately, they are not able to meet their son and start working in the quarry, where Nathan meets with his death. Rukmani comes back to the village to live with her son Selvam and daughter Irawaddy.

2.2.1 The impact of tannery on agrarian culture

Nectar in a Sieve (2009) exhibits the modern developmental attitude in the form of tannery that encroaches on the cultivated land and the age-old agrarian culture of the village. Markandaya tries to link the introduction of tannery in the village to two groups of men of different provenance. The Zamindari is a system under which the landholders collect tax from peasants whom they have given land for cultivation; this is represented by Sivaji, whereas the Savakari, business of money lending, is by the moneylender Biswas and the grocer Hanumant.

The construction of tannery starts under the supervision of an overseer and white men. Markandaya's purpose is to depict the idyllic life of the village that is desecrated by the introduction of tannery. Before the introduction of tannery, the village was calm and serene with bountiful flora and fauna, and the major occupation of the villagers was agriculture. Over the period of two months, due to the construction of tannery, the entire picture of the village is completely changed. There is a continuous overflow of bullock-carts laden with bricks, stones, cement, sheets of tin, corrugated iron, coils of rope and hemp, which disrupts the surrounding environment. Further, the kilns in neighbouring villages that keep burning the bricks are unable to meet the demand of construction of tannery. As the young boy Arjun aptly expresses, "They are pulling down houses around the maiden and there is a long line of bullock carts carrying bricks" (Markandaya 2009:27). It builds an ambience of danger lurking in the form of global warming presented through nature's fury, which comes in the shape of either draught or flood. The evil of development has not only disturbed the agrarian culture but also forced villagers to work on the project of construction of tannery.

The second important feature is represented by the landowning Zamindari, portrayed in the novel through Sivaji, the proprietor who leases his land to peasants, Rukmani and Nathan, who are the tillers of the Sivaji's land. However, Sivaji is least bothered about the livelihood of Nathan and Rukmani, who have worked on the land for more than 30 years. Regardless of more than 30 years' association with the land as if it is their own child, everything that Rukmani and Nathan held is effaced by Sivaji.

The landowner appears one day and declares that the land is to be sold to the tannery and has to be vacated within two weeks. This inhuman attitude is received as follows by Nathan, "The land is to be sold. We are to move. Sivaji came this morning. He says there is nothing to be done. The tannery owners are buying the land. They pay good prices" (134).

Nevertheless, the introduction of tannery affects the villagers' lives at different levels. First, bazaar prices of daily commodities soar out of the reach of the common people. The moneylenders – Hanumanta and Biswas – exploit the villagers during their hard times. Indifferent to the villagers, the shopkeeper Hanumanta is very harsh towards Rukmani with his words, "You have come for rice. They all come for rice. I have none to sell, only enough for my wife and children. . . . Are you not growers of it? Why then do you come to me?" (45). This very statement of Hanumanta shows the deteriorating condition of agriculture. Then, Rukmani goes to Biswas to buy rice; he too is very apathetic to the villagers. He fleeces the villagers and says, "Take it or leave it. I can get double that sum from the tanners, but because I know you" (45). This attitude shows how he is unconcerned about villagers' plight. We become aware of the fact that tannery has changed everyone's attitude in the village.

Second, the small-scale businessmen are wiped out because of bigger shops. For instance, Rukmani expresses her dissatisfaction over the impact of tannery on an aged woman's business of selling vegetables. Earlier, Rukmani used to sell vegetables to granny, but as bazaar prices have gone up too high she starts selling vegetables to the shopkeeper, Biswas, in order to make more profit. However, this has an adverse impact on the old woman's business. Here, Markandaya has given a starkly contrasting picture of the village before and after the introduction of tannery, which brings a paradigmatic shift in the villagers' lives, "But we never went hungry as some families were doing. We grew our own plantains and coconuts, the harvests were good and there was always food in the house – at least a bagful of rice, a little dhal, if no more" (26). Therefore, the agrarian culture and the small-scale shopkeepers' condition steadily and gradually deteriorates by the slow sprawl and spread of the flourishing tentacles of tannery, which in the end swallows the serene pastoral land of the village.

Third, tannery changes the attitude of the young generation, as they no longer want to continue their ancestral occupation of tilling the land; rather, they wish to earn easy money by working in tannery. Rukmani's sons – Arjun, Thambi, and Raja – have not shown any inclination to cultivate the land; instead, they would like to join tannery. In response to Arjun's decision to join tannery, Rukmani expresses her contention and remorse in these words, "You are young, besides, you are not of the caste of tanners. What will our relations say?" (53). Here, we can observe the recklessness and indifferent attitude of Arjun in his reply, "I do not know, I do not care. The important thing is to eat" (53). Tannery not only engulfs the agrarian culture but also creates a fractured identity among the younger generation. Owing to insufficient food, first Arjun and next Thambi, two of Rukmani's sons, join the tannery and later go to Ceylon to work on plantations. Still worse, Raja – another of Rukmani's sons who joined tannery – is caught in a theft in tannery and loses his life. This adds more to the woes of Rukmani's already devastated life.

Fourth, the tannery denotes the indirect effect in the form of global warming on the age-old agrarian profession and the environment. The villagers struggle hard to survive. The fields have consumed all their labour, and all that lies before them in the end are worthless heaps of dried hay on account of nature's fury. The condition of landless people is pathetic, as they receive no concessions in paying their dues to their landowner and are left with nothing; their only hope is to wait for another crop. Here, it is very appropriate to relate Carolyn Merchant's confrontation of "technology innovation, the spread of the capitalist market, *the scientific revolution*, and changing attitudes towards nature and the earth" (Merchant 1980:43; emphasis added). Indeed, once peaceful and calm, Rukmani's life changes into that of hunger and suffering:

> The drought continued until we lost control of the time. Day after day the pitiless sun blazed down, scorching whatever still struggled to grow and baking the earth hard until at last it split and irregular fissures gaped in the land. Plants died and the grasses rotted, cattle and sheep crept to the river that was no more and perished there for lack of water, lizards and squirrels lay prone and gasping in the blistering sunlight.
>
> (Markandaya 2009:79)

The novel, further, projects the ugly transition of the simple, traditional, pastoral and idyllic village into a crowded noisy town. The birth of a town in the village entirely changes the scene of the village and the lifestyle of the villagers, who just experience destruction, frustration and long-lasting poverty:

> Not in the town, where all that was natural had long been sacrificed, but on its outskirts, one could still see the passing of the seasons. For in the town there were the crowds, and streets battened down upon the earth, and the filth that men had put upon it; and one walked with care for what might lie beneath one's feet or threaten from before or behind; and in this preoccupation forgot to look at the sun or the stars, or even to observe they had changed their setting in the sky: and knew nothing of the passage of time save in dry frenzy, by looking at a clock.
>
> (117)

Fifth, owing to hunger and loss of traditional modes of work, many of the peasant women are compelled to take up prostitution. Kunthi, Rukmani's neighbour, starts the business of prostitution to fulfil her hunger as well as that of her family. Markandaya very minutely extrapolates, "I thought of Kunthi as I had once seen her, with painted mouth and scented thighs that had held so many men, and I wondered if after all these years he had not at last found about her. Perhaps the truth has been forced upon him, I thought, looking at her with suspension, and I gazed upon that ravaged beauty" (84).

The throng of men has spoiled women's chastity. Not only Kunthi but even Ira – daughter of Rukmani – who is abandoned by her husband because of her barrenness, could not bear her grim, dull, dark, hopeless future, the unending hunger and starvation of her younger brother, and slowly turns her mind to prostitution. In this context, Rukmani's thinking is very apropos when she says:

> But the man who finds a woman in the street, raises an eyebrow and snaps his fingers so that she follows him, throws her a few coins that he may possess her, holds her unresisting whatever he does to her, for this is what he has paid for – what cares such a man for the woman who is his for a brief moment? He has gained his relief, she her payment, he merges carelessly into the human throng, consigning her back into the shadows where she worked or to the gaudy streets where she loitered.
>
> (118)

Once Ira was a decent and obedient girl but now she is ready to sell her body at the cost of one rupee per day in order to save her brother Kuti, who is lying hopelessly on his deathbed. Tannery has not only marred the village pastoral land and agrarian culture but also ruptured moral values.

2.2.2 *Woman as a saviour and a nurturer of nature*

The novel confirms the viewpoints of most of the well-known ecofeminists that women are invested with a mission to save and nurture nature. Chris Cuomo, an American ecofeminist, observes, "Environmental ethics can benefit by incorporating feminist insights on the limitations of traditional, philosophical conceptions of ethics" (Cuomo 1988:i). This key feature of ecofeminism can be studied through Rukmani – the principal protagonist of the novel.

Markandaya employs the first-person narrative technique to show the gravity of the problem of introduction of tannery in the village and its impact on Third World peasant women. The protagonist Rukmani marries Nathan, a tenant farmer, poor in a material sense but rich in love and care. Rukmani is strongly associated with nature; she nurtures and cares the field like her own child. The tannery – in her experience – is a catastrophe that falls upon the village, not only disturbs the simple, primitive, traditional and agrarian-oriented families, but also the pastoral land of the village.

Rukmani's character can be studied at two levels: first, her association with nature, and second, her reaction to the introduction of tannery, which affects the village pastoral land and the lives of the villagers. Both reflect ecofeminist concerns. After marrying Nathan, Rukmani starts her journey in a bullock cart to her husband's village. Her concern for animals is narrated in the following manner, "Poor beasts, they seemed glad of the water, for already their hides were dusty" (Markandaya 2009:5). She likes the song of mynahs and the cry of the eagle, that of many other birds, which makes her warm and drowsy. Rukmani takes pride in planting seeds and nurturing plants in the garden. She plants a few pumpkin seeds in the garden behind the hut and soon the seeds sprout with delicate green shoots. She frequently visits the nearby well to fetch water for the plants. Sometime later, a pumpkin begins to ripen into yellow and red. She has a lot of admiration for it: "One would have thought you had never seen a pumpkin before" (11). The growth of this pumpkin boosts her energy and she starts planting beans, sweet potatoes, brinjals and chillies. She is certain that all these plants grow well in her hand.

Rukmani's intimate and intricate relationship with nature is portrayed through her labour in the fields, which represents her affection for nature. Indeed, Rukmani prays continuously for the betterment of her land and crops, fruits and harvest. She shows a divine integrity between herself and nature upon recounting, "I was young and fanciful then, and it seemed to me not that they grew as I did, unconsciously, but that each of the dry, hard pellets I held in my palm had within it the very secret of life itself, curled tightly within, under leaf after protective leaf" (14).

As an Indian ecofeminist Vandana Shiva posits,

> For more than forty centuries, Third World peasants, often predominantly women, have innovated in agriculture. Crops have crossed continents, crop varieties have been improved, and patterns of rotational and mixed cropping have been evolved to match the needs of the crop community and the ecosystem. . . . Peasants as experts, as plant breeders, as soil scientists, as water managers, have kept the world fed all these centuries.
>
> (Shiva 2010:98)

Markandaya focuses on the Third World women peasants and sheds light on their attitude towards variegated cropping and the maintenance of sustainable ecological balance. Rukmani and her husband stress the importance of mixed cropping in paddy. Rukmani – the Third World woman peasant – has traditional agrarian knowledge of cultivating the land, "Dung is too useful in our homes to be given to the land, for it is fuel to us and protection against damp and heat and even ants and mice. Did you not know?" (Markandaya 2009:34). Dung is indeed one of the major fertilizers, germicides and fuel for Indian villagers.

Though she is not able to work in the field because of her pregnancy, she starts working in the garden. The growth of plants and vegetables keeps her in constant wonder. Her close association and affinity with the land is increased by her physical, emotional, sexual and psychological development through her work in the garden and echoes spiritual ecofeminist ideology. The interconnection between woman and nature can here be traced through activities of Rukmani related to the field. This lineage of feminist spirituality is advocated by the American ecofeminist Starhawk, who claims "the second base concept of earth-centred spirituality is that of interconnection . . . (this) translates, natural cycles and processes, animals and plants" (Starhawk 1989:174). Due to nature's fury, Rukmani's family lost crops many times and went empty stomach many nights, yet she maintains her balance and never curses either field or nature because of her faith in the motherland as the following passage in the novel indicates:

> While the sun shines on you the fields are green and beautiful to the eye, and your husband sees beauty in you which no one has seen before, and you have a good store of grain laid away for hard times, a roof over you and a sweet stirring in your body, what more can a woman ask for? My heart sang and my feet were light as I went about my work, getting up at sunrise and going to sleep content. Peace and quiet were ours. How well I recall it, how grateful I am that not all the clamour which invaded our lives later could subdue the memory or still longing for it. Rather, it has strengthened it: had there not been what has been, I might never have known how blessed we were.
>
> (Markandaya 2009:9)

Vandana Shiva thinks that the exploitation of women and nature is due to development attitudes in the form of scientific revolution and a reductionist paradigm which she equates with mal-development: "Development has meant the ecological and cultural rupture of bonds with nature, and within society, it has meant the transformation of organic communities into groups of uprooted and alienated individuals searching for abstract identities" (Mies & Shiva 1993:99).

Rukmani's compassion for nature is visible in her apprehension for the bullock, which has developed many raw patches on its skin with a trickle of blood running down. It leaves a deep scar on her heart. But the cart driver mutters indifferently that the animal will soon be good for nothing and he is not able to afford another one. These contrasting emotional responses clearly depict, as the major ecofeminists contest, that women have greater concern for animals whereas men tend to be more indifferent to them. A similar observation can be made when Rukmani accidently touches a snake, which Nathan, her husband, kills. In this context, Markandaya says, "Women can sometimes be more soothing than men" (Markandaya 2009:17). The novel features women as harmonious with nature, tending to follow traditional Indian cultural customs like the belief that a cobra is sacred and should be worshipped instead of killed. Moreover, she names her daughter after one of the rivers of Asia, Irawaddy, as water is considered a precious and sacred resource.

Rukmani's reaction to the introduction of tannery in the village is noteworthy. She regrets that tannery has invaded the village and the maiden land where children used to play. The end result of tannery is that small farmers generally lose their livelihood because their sons are lured off the land by the paid work. Rukmani and her husband can no longer pay their dues, leading the landowner to sell the land to the tannery. Rukmani's phobia of modernisation is clearly visible in these lines, "the tannery would eventually be our undoing. . . . Since then it had spread like weeds in an untended garden, strangling whatever life grew in its way" (Markandaya 2009:135–136).

In addition to loss of crops and having to till the land and sow the seeds, she sells her wedding sari, her daughter's sari and Nathan's dhoti for the sake of land. This shows that Rukmani is not ready to accept neither modernisation nor the introduction of tannery in the village. She is forced to confront noise and smell of tannery as well as Ira's adultery and her gloomy future. On the one hand, Rukmani loses her sons because of the modern juggernaut (the tannery): Arjun and Thambi leave their ancestral occupation of agriculture and join tannery, later leaving for Ceylon to work on a plantation; Raja, another son, loses his life in the tannery; and Selvem finds work in the city. Rukmani then witnesses the adultery of her daughter and her cruel fate: when her son-in-law had desired a child, he had not been given one, but now she is blessed with an illegitimate grandchild. On the other hand, tannery leads to draught. The impact of draught on Rukmani's family is horrific. They become penniless, as all the money has gone to pay the land dues and nothing is left to sell; everything withers in the long weeds of draught including the crops of paddy, vine and vegetables. In short, the family of the Rukmani withers in the whirlwind of tannery. It forces them to leave the village.

Rukmani's family meets with a tragic end. Being landless in spite of working on the land for more than 30 years, Rukmani and Nathan end up going to the city to live with their sons. However, they are unfortunately unable to meet them. With no money to return to the village, they start working in a stone quarry, where Rukmani loses her husband. The novel is a rigmarole journey of a Third World woman peasant, impelled to move from a simple, naïve and traditional agrarian context to the city for survival. Along with these core elements of ecofeminism, we will examine Markandaya's concern for environmental problems.

2.2.3 Commercial culture vs. pastoral culture

Markandaya draws our attention not only to the pathetic condition of Rukmani's family but also projects the impact of tannery on the pastoral land. The introduction of tannery in the village epitomises commercial culture. In this section, we will discuss how tannery not only grabs the livelihood of people, but also mars the ecology and environment of the village. It also transits the silent village into clamours one. The pastoral land is encroached upon by the construction of a little colony for higher officers and workers, "brick cottages with whitewashed walls and red-tiled roofs" (Markandaya 2009:49–50). Here, Markandaya probes into two contrasting lifestyles: on the one hand, the construction of concrete buildings in between the village and the open country, on the other hand, villagers' typical huts with thatched roofs and mud walls. The commercial culture replaces the natural serene land: "A large building, spruce and white; not only money has built it but men's hopes and pity, as I know who have seen it grow brick by brick and year by year" (3). The natural eco-friendly houses are pulled down and a concrete jungle starts taking shape. Once the village was clean and unpolluted, and the villagers were able to grow crops to fulfil their daily needs and lead a happy life. But, due to the introduction of tannery, land, air and water are polluted to such an extent that villagers face the problem of draught. The adverse impact of tannery can be sensed through the gradual disappearance of birds from the village. The following excerpt shows the biodiversity of the land:

> The air was cool and still, yet the paddy caught what little movement there was, leaning slightly one way and the next with soft whispering. At one time there had been kingfishers here, flashing between the young shoots for our fish; and paddy birds; and sometimes, in the shallower the water reaches the river, flamingoes, striding with ungainly precision among the water reeds, with plumage of a glory not of this earth. Now birds came no more, for the tannery lay close – except crows and kites and such scavenging birds, eager for the town's offal, or sometimes a palpitta, skimming past with raucous cry but never stopping, perhaps dropping a blue-black feather in flight to delight the children.
>
> (71)

Markandaya espouses man's indifferent attitude towards animal: "A never-ending line of carts brought the raw materials in – thousands of skins, goat, calf, lizards and snake skins – and took them away again tanned, dyed and finished" (49). Initially, it was very difficult for the villagers to cope with tannery because it intrudes with the serenity of the pastoral: "I had got used to the noise and smell of the tannery; they no longer affected me. I had seen the slow, calm beauty of our village wilt in the blast from the town, and I grieved no more" (64).

Owing to the introduction of tannery, there is a drastic change in seasons and natural phenomena, which further introduces global warming: "Each day the level of water dropped and the heads of the paddy hung lower. The river had shrunk to a trickle, the well was as dry as a bone. Before long the shoots of the paddy were tipped with brown; even as we watched, the stain spread like some terrible disease, choking out the green that meant life to us" (74). These are a few consequences of the impact of tannery on nature and human beings.

2.2.4 Agriculture: means of subsistence

For Indian peasants and tribal people, agriculture is a means of subsistence and a question of survival. Nathan comes to know that if he is not able to pay the dues in time, he will lose the land. This idea of losing the land disturbs him greatly; he is not able to think of the consequence of this possibility. However, his impeccable faith in the land makes him withstand all the woes and worries. Additionally, the reaping of a meagre crop forces Nathan and Rukmani to sell their utensils and clothes. Despite losing their paddy due to nature's harsh and unbearable fury, Nathan and Rukmani do not lose their hopes and aspirations and Nathan starts breaking the dams in the fields where water is filled. Attributed to the effects of draught or floods, Nathan loses his crops. For many days the entire family starves.

The eternally optimistic, Nathan never falls prey to adverse circumstances. Nathan believes that the cycles of nature change and it will be a matter of time for a good harvest to be reaped. With the turn of the tide, they may be able to pay the dues to the landowner, Sivaji, and even manage to visit their son who works in the city.

Nathan has profound connection with the land. Though the land is not his own, he loves the land as if it is his own child not just because it is the means of subsistence but he has very intimate bonding with it. However, when he comes to know that his land is going to be sold to the tannery builders, he does not want to join the available modern means of production (i.e., the tannery) for survival. This shows man's affinity with nature. He always harbours an ardent desire that he will have his own land someday because it will be a permanent source of income: "He had always wanted to own his own land, though the years had been the hope, growing fainter with each year, each child, that one day he would be able to call a small portion of land his own. Now even his sons knew it would never be" (Markandaya 2009:54). This is an example of how men are also implicated in the crisis of environment.

2.2.5 Rhetorical tropes

An interesting aspect of the novel is the graphical narration of flora and fauna. Markandaya creates the nature's ambience to such an extent that the readers feel as if they are really moving in the paddy fields enjoying nature's scenic beauty. The reader's journey begins with green paddy fields and progresses to the twittering of sparrows, notes of mynah along with many more birds and even accompanies the brook that runs near the field. The entire novel and each chapter open with the scenic beauty of nature. Nature is one of the major characters, narrating its own tale of woes. Through this we can get a picture of the Indian agrarian culture and biodiversity of various species and birds. The pastoral land is the primary setting and the entire novel is woven around it. The basic purpose of this unique narration is to sketch the Indian natural beauty, the role of nature in Indian culture and the impact of developmental attitude in the form of the tannery on the rustic life. Most of the chapters open with natural beauty and natural phenomena that include other aspects also, for example, it gives a glimpse of our indifference to nature and the exploitation of nature and its consequences. It is said, "Nature is like a wild animal that you have trained to work for you" (Markandaya 2009:41). But, this remark also goes to the same line, "As soon as the rain was over, and the cracks in the earth had healed" (81).

On the other hand, the unique narrative technique personifies the impact of tannery through the transition in the river and the paddy fields: "Each day the level of the water dropped and the heads of the paddy hung lower. The river had shrunk to a trickle, the well was as dry as a bone" (74).

2.2.6 The uniqueness of **Nectar in a Sieve** *as an ecohumanist narrative*

The book, after examining thoroughly the treatment of ecofeminism in this novel, tries to ascertain whether the select Indian novels add any new dimension to the ecofeminist discourse.

The major ecofeminists posit that women are saviours of nature and vanguards to protect, preserve and nurture nature (D'Eaubonne 1974; Ruether 1975; Daly 1978; Merchant 1980; Mies & Shiva 1993). However, the given reading of *Nectar in a Sieve* shows that how Indian novels add a new dimension to the ecofeminist discourse, indicating that men also are saviours, nurturers and caretakers of nature as well as subjugated by development attitude.

Nathan always wishes that his sons should work on land and follow the agrarian culture for which he has a divine passion and attachment. He never supports modern developmental attitude in the form of tannery. Therefore, he convinces his sons to work on the field but his efforts fail. His sons turn their back to the land and say, "There is nothing for us here, for we have neither the means to buy land nor to rent it. Would you have us wasting our youth chafing against things we cannot change?" (Markandaya 2009:70)

Despite enduring many difficulties, the consequence of nature's fury and loss of his sons, Nathan manages to face these hardships and maintains faith in the land. For many days Nathan and his family go empty stomach because of draught or flood. However, he never gives a second thought to work in the tannery to get rid of hunger and starvation. This aspect comes across in the following excerpt, "We will find our strength. One look at the swelling grain will be enough to renew our vigour. Indeed, it did our hearts good to see the paddy ripen. We watched it as a dog watches a bone, jealously, lest it be snatched away; or as a mother her child, with pride and affection. And most of all fear" (Markandaya 2009:96). Moreover, when he comes to know that Sivaji sells the land to tannery for a good price, Nathan is not able to cope with the situation. His hands tremble and he feels impotent and is unable to say anything because he thinks that he may lose affinity, intimacy and dependency on land. He concludes that the tannery has resources beyond his imagination, ensuring a good harvest and crop of rice to feed tannery workers.

Nathan's concern, affinity and oneness with nature make us ponder over the conceptual understanding of the spiritual ecofeminism that is the interconnection between women and nature; the novel shows man and nature's interconnection. It adds a new dimension to the ecofeminist discourse. Nathan engages in either cultivating the land or looking after the animals – bullocks and goats. When Nathan sees a ripened pumpkin fruit – his enthusiasm is at its height and he is full of admiration. His admiration, involvement and proximity to the field symbolise his concern for nature. Nathan feels a great joy, happiness, passion, devotion and dedication for the land. He is always preoccupied with either the land or the crops. Trying to show the growth of crops, he brings a few green stems to Rukmani, "Look at our land – is it not beautiful? The fields are green and the grain is ripening. It will be a good harvest year, there will be plenty. . . . See how firm and strong they are – no sign of disease at all. And look, the grain is already forming" (Markandaya 2009:71). The minute and detailed explanation of the growth of grain reflects the interconnection between man and nature.

The joint venture of Nathan and Rukmani projects that Indian agriculture is gender neutral and cultivation of land, ploughing, sowing and harvesting the crop needs the involvement of men and women. As Rukmani proffers, "Sowing time was at hand, and I was out all the day with Nathan planting the paddy in the drained fields. Corn had to be sown too; the land was ready. My husband ploughed it, steadying the plough behind the two bullocks while I came behind, strewing the seed to either side sprinkling the earth over from the basket at my hip" (Markandaya 2009:18). Nathan and Rukmani live their lives for the sake of their land as though it is their own child as well as considering it as part of their life to such an extent that they are not prepared to leave their land even in the moment of crisis. They never think of the land merely as a means of survival. Rather, they feel a spiritual connection with it. As Starhawk writes, "We walk on has a scared character, we cannot allow its soil to become eroded. We cannot clear-cut the old growth forests for profit, for what is sacred cannot be measured against the scales of profit and loss" (Foss et al. 1999:177).

Thus, Nathan's propinquity to the land gives an opportunity to contemplate and rethink that male protagonist is also connected to nature. As such, we can see that Markandaya's novel allows for a critic of the major ecofeminist conceptual argument, which surfaces in Rosemary Radford Ruether's proposition that is "patriarchal self-deception about the origin of consciousness ends logically in the destruction of the earth" (Ruether 1975:195). This argument reveals that the basic contention of ecofeminism is not universal in scope, but rather the outcome of a certain culture and it can be opposed from another vantage point.

We can rethink two other minor characters of the novel, Sivaji and Kenny, in this line of thought. Sivaji is a landowner and in the end he sells his land to tannery. However, the entire novel projects that he is always with Nathan's family in their thick and thin and gives them time for the payment of the dues. On the other hand, Kenny, the white man as a doctor, neither relates to nature nor tries to be a saviour of nature directly but has concern for the villagers and supports them in their endeavours.

Several ecofeminists argue capitalist patriarchy subjugates women and degrade nature. However, the select Indian novels add a new dimension to the ecofeminist theory that due to developmental attitude, man is also subjugated. Nathan, in *Nectar in a Sieve* (2009), loses his means of livelihood. Therefore, there is no hope for him to live in the village. Both Nathan and Rukmani set out for a new life in the city where they can stay with their son. In the cart, on the way to the city after winding up their 30 years of life in a small bundle of luggage, Rukmani's and Nathan's anguish is very acutely projected in the following manner, "The hut – its inhabitant – recedes behind us and yet in front of us, for we are sitting with our backs to the bullocks. Our beloved green fields fall away to a blur, the hut becomes a smudge on the horizon. Still we strain our eyes to pierce the reddish dust the wheels throw up" (Markandaya 2009:144).

Thus, despite having spent their entire life working very hard on the land, Rukmani and Nathan are left with nothing in the end. Being forced to join a stone quarry in the city, Nathan meets his tragic death there. Nathan is thus a first-hand victim of the introduction of tannery in the village.

Therefore, the emergence of tannery in the setting of the village changes the face of the village beyond recognition besides altering the lives of many people. Although many are able to survive successfully, many more fall victims to tannery and lose their lives in the clutches of this modern juggernaut. Tannery brings only resentment and resignation to the lives of villagers because their sons and daughters are allured by tannery; ultimately they lose their rural lifestyle. Thus, after reading Markandaya's ecofeminist narration, let us move to yet another significant novel, Pundalik Naik's *The Upheaval* (2002).

2.3 *The Upheaval*: the impact of mining on farming community

Pundalik Naik's *The Upheaval* 2002 is the first Konkani novel to be translated into English by Vidya Pai. Keeping his own village as a canvas for the novel along with his personal ethos of rustic and agrarian life, Naik in *The Upheaval* (2002) visualises how mining has drained not only the daily life of the village folk but also their culture, tradition, festivals and seasonal activities. The novel gives an optical image of a major environmental crisis.

Pundalik Naik is one of the most outstanding pioneering dramatists and prolific writers from Goa. He is accredited with 32 plays, several collections of short stories, novels, novellas and two films. Though he has dealt with many genres, his predominant contribution is to the Konkani theatre. Most of his plays narrate the plight of lower classes – a dark social theme – and the impact of mines on the lives of the village folk and their agrarian culture. His writings project environmental issues and make a mockery of political and administrative system, corruption and injustice. His major plays include *Ransundari* (1974), *Chhappan Thigali Yashwantrao* (1980), *Suring* (1982), *Demand* (1986) and *Shabai Shabai Balmjan Samaj* (1986). He was bestowed with the Sahitya Akademi Award in 1982.

The Upheaval (2002) espouses the impact of mining on the river Mandovi, the land of Kolamba and the agrarian culture of the village. On the one hand, Abu, the village old man, persistently strives to guard the village Kolamba and the river Mandovi from mining. On the other hand, Naik discerns how Pandhari, the protagonist, becomes a prey to mining. Earlier, he led a very simple peasant life along with his daughter Kesar, son Nanu and his wife Rukmini in the village Kolamba, which is situated on the bank of the river Mandovi. He is always engrossed in farming work. But, the evil in the form of mining, which gradually spreads its tentacles in Kolamba, enters his life. The village man Babuso and other Gujarathi gentleman entice Pandhari, who represents the peasant community, to join mining. Earlier, Pandhari was very proud of his field and harvest. Meanwhile, however, he leaves behind his age-old agrarian culture and joins mining by ignoring his wife Rukmini's plea not to join mining. Naik would like to project that mining has not only affected the environment but also the age-old serene culture and sanctity of women. Rukmini, a simple, down-to-earth and family-oriented woman who is always busy with domestic and field chores, shifts her attitude to crookedness, decisiveness and falls prey to Babuso's carnal desire. But having lost her health, she becomes bed-ridden. Kesar works very hard on mining and she too loses her chastity. Later, she gets pregnant and runs away.

On the other hand, mining makes Nanu leave school and join mining. He falls prey to all bad habits and involves in liquor and women. In the end, while working in the mines he meets with his death. When Pandhari works on the fields, he is rich and happy, but when he joins mining; he loses his land. He is not even able to work on the mining because he contracts Tuberculosis. To survive, he starts business as a tea vendor. Moreover, mining pollutes the land of the village Kolamba and contaminates the river Mandovi.

2.3.1 Mining industry in Goa

Pundalik Naik was born and raised in a farming community with sowing, ploughing and harvesting "a process that is both exhausting and exhilarating" (Naik 2002:xxvi). Pundalik Naik's reminiscences form the backdrop to the novel. He says,

> I belonged to a family of farmers and have experienced poverty and suffering. As a boy, besides helping the elders in the field, I had to look after the

> cattle. Wonderful surroundings – hills and rivers – had a profound impact on my mind. Nature inspired me and instilled in me a competitive spirit. I had a lot to say and started it with poem.
>
> (xxvi)

We need to examine the iron ore mining industry in Goa to explore the thematic concern of the novel *The Upheaval* (2002). The Portuguese regime understood that a lot of iron and steel is needed to rebuild the ruined Japan; and they had this advantage in Goa.

Consequently, hundreds of hectares of forestlands have been literally destroyed by mining. It has not only disrupted the agrarian culture but also disturbed the farming community. The largest 30 mines are situated in the catchment area of the Mandovi River. The Zuari River, too, has been highly silted with mining waste. An environmentalist Ramesh Gauns articulates, "These two rivers are Goa's lifelines, accounting for 67 per cent of the state's freshwater sources as well as mining contributes just 6 per cent of Goa's GDP – the same as agriculture" (Takle 2012:37).

Further, Takle says that about 400 million tons of iron ore has been extracted from Goa and hardly 300 million tons remain. At the rate of 50–60 million tons a year, mining can go on for a maximum of 10 years. He said that after the extraction of the ores the companies would pack up their bags. Here, Takle poses an implicit question, "And the people of Goa?" This question perpetuates the intensity and the long-lasting effect of mining on the livelihood of the local people and the surrounding land and river (Takle 2012).

The core purpose of this research is not to give any statistical analysis of mining in Goa but to investigate the writer's concern for environmental problems and its impact on the peasant community and nature. Let us examine the ecofeminist concern in *The Upheaval* (2002) by evaluating the characters and their journey from the simple rustic farming occupation to the mining mafia.

2.3.2 Modern developmental attitude: mining

Pundalik Naik explores the first intervention of the ecofeminist discourse, namely, the modern developmental attitude in the form of mining. This is tracked in the novel through certain characters: Pandhari, the protagonist of the novel; Prasad Babu, a Gujarati man; Babuso, a villager and Nanu, the son of Pandhari.

Let us begin our analysis by understanding two minor but the most important characters of the novel – Babuso and Prasad Babu – because of whom the mining fever emerges in the village. Prasad Babu, a Bengali man, wants to induce the people of the village to join the mining work with the help of Babuso. Prasad Babu's physical appearance represents developmental attitude and Babuso's appearance projects the agrarian way of life. Naik shows disparities between two cultures and lifestyles – simple agrarian culture/modern developmental attitude. Prasad Babu is an epitome of the modern attitude, whereas Babuso is the embodiment of agrarian culture:

> A waxy black moustache shaped like the petals of a water-lily. Gold buttons on a shirt that hung to his knees. A thick wallet and a couple of pencils stuck out of his pocket. He wore a pair of sandals with thick soles with fine straps. . . . He sat on the clay seat looking like the snow-white cranes that perched on the embankments in the fields. And like the cranes, he turned his gaze this way and that. Sitting beside him in his shabby, worn cloths, Babuso resembled nothing more than a crow.
>
> (Naik 2002:11–12)

Though Naik has portrayed Prasad Babu's role as a passing reference on the canvas of the novel, he is the one who leaves an irreparable scar on the agrarian culture and farming community. Owing to Prasad Babu's and Babuso's alluring words, the people of Kolamba gradually start to leave behind their age-old occupation of farming to join the mining work which is an easy source of earning. However, they never give a second thought to the long-term effect of mining on themselves as well as nature.

Babuso, being cunning and crooked, earns money for his livelihood by enticing villagers to join mining. In spite of knowing the importance of farming in the villagers' lives, he is overshadowed by money and greed as Naik sketches, "They're not just any type of stone. . . . they say they make gold out of them. And that pit is called a mine. There's lots of work up there. You sit here doing nothing for six months after sowing your field. It's possible that you won't get a good crop. Even if you do, it'll only be because the cows and other animals don't get to it first" (12). He is the one who has disrupted the chastity of women in the village and coaxed women to join mining. Naik represents how Indian agrarian culture is transformed into commercial culture and how there is a dual war between the agrarian culture and the modern developmental culture.

Prasad Babu and Babuso's visit in the pitch-dark evening to Pandhari's house to ask him to join mining holds a symbolic meaning. The word 'dark' is a symbolic gesture of destruction, which is hovering over the village in the form of mining. The mining mafia slowly surges into the village through these two characters. Babuso tries to convince Pandhari as per Prasad Babu's offering to join mining to earn money. Moreover, he tells him that if he wishes to join with his bullock-cart, he can earn more money, 70 rupees, for each trip. Just his bullock-cart has to bring stone chips from the mine, Naveli, to the river Shenori where the mining people load stone chips in big ships for export. In spite of a lot of opposition from his wife and many villagers, Pandhari joins mining along with his cart by which he can get double income. In this way, the protagonist of the novel is first prey of the mining project.

Naik's basic purpose behind narrating the story of Pandhari is to capture the catastrophic consequences of mining mafia on the villagers' lives and on the serene natural land. Through this narration, he wants to sustain and register the impact of mining on the minds of readers. This signature tone runs throughout the novel to depict two counter-cultures that are trying to exist together. Pandhari prepares his land for sowing after tilling, ploughing and hoeing with a lot of hard work. However, in the meantime, the evil in the form of mining is introduced in the life of Pandhari. Pandhari is tempted by the words of Babuso and joins mining along with his bullock cart to carry stones. He stakes his ancestral occupation of farming.

On the other hand, his wife Rukmini is least bothered about mining work and engrosses herself in the preparation of sowing. When Rukmini opposes Pandhari's proposal of joining the mining work along with his bullock cart, he hits her very hard. This reaction of Pandhari to Rukmini is not because of intoxication caused by alcohol. His infatuation for mining has worked on his mind. The temptation of earning more money by working in mines instead of working on the fields is an emblematic representation of man's contemporary attitude. It replaces the traditional agrarian and cultural values. Here, Naik shows that mining is the developmental attitude, which is the symbol of patriarchy, and Pandhari falls prey to it. The Indian ecofeminist Vandana Shiva asserts:

> Dams, mines, energy plants, military bases – these are the temples of the new religion called 'development,' a religion that provides the rational for the modernizing state, its bureaucracies and technocracies. What is sacrificed at the altar of this religion is nature's life and people's life. The sacraments of development are made of the ruins and desecration of other sacred, especially sacred soils. They are based on the dismantling of society and community, on the uprooting of people and cultures.
>
> (Mies and Shiva 1993:98)

Always Pandhari is the first man in the village to finish ploughing and the first one to reap the harvest and to get a good crop. However, that year he is not able to reap a good harvest because of working in mines. He fails to tend his field properly. This is the first onslaught on the agrarian culture. In this context, we can see the patriarchal attitude of Babuso, when he says, "But that's hard work in the sun. And you hardly get any paddy at the end of the day. Why don't you take her to the mine at Naveli?" (Naik 2002:42) Babuso thinks that there is nothing left in the fields. Even, he asks Pandhari to make Rukmini, his wife, join mining by which both of them can earn more money.

Pandhari's indifference to the fields and his family is an astute illustration of the transition of agrarian culture (the age-old tradition) into mining (the commercial culture). The dialogue of the schoolteacher Salvo Master carries the crux of the issue: "You only understand the language of money these days. Not the language of men" (Naik 2002:74). Pandhari gives up his age-old occupation for the sake of mining, but later he loses mining work too for two reasons. First, due to his ill health he is not able to work in mines. This is the result of the continuous carrying of loads of chips and pollution by which he contracts tuberculosis. Second, he has lost his bullocks. In addition to this, the introduction of trucks, barges, big machines with shovels in place of bullock-carts, and the jetting with moving belts that carry the ore right into the barge has grabbed the work of the workers in the mine. The business of carrying the stones has ended. Here, we may infer that commercial culture may make a lot of money for us but it is temporary, whereas agriculture may not yield like mining but it is a permanent source of income.

The evil effects of mining not only affect one generation but they have taken a toll on many generations to come. Naik wants to expand our understanding of the severity of problem through the decision of Pandhari to join mining. It has a long-lasting impact on his son Nanu, who represents the young generation of the village. Nanu is very much obsessed with the activity of sowing in the field. On the occasion of the celebration of sowing, he invites his master for lunch. Due to his father's decision to join mining, the activity of sowing is postponed which disturbs Nanu to such an extent that he is unable to concentrate on the mathematics class. When the teacher asks him an arithmetic sum like – suppose a man earns 2 rupees per day, the immediate response without his knowledge, Nanu says that 5 rupees – a reflection of his father's conversation with Babuso – and as the master says 10 rupees for the cart, then he adds that 70 rupees for the cart and at the end of the class he bursts into tears. This enumerates the young generation's nostalgic roots in their age-old occupation. Nanu wants to go to school and continue his education and simultaneously work on the field, but his father's pressure to join mining makes him quit school and also the field.

Naik attempts to explore how the young generation, without understanding the consequences, inculcates the actions and activities of their parents. During the ritual *'dhalo'* the children Nanu, Kesar, Narshinv and Shankar gather on the banks of the river. They want to play some game. Suddenly, Shankar says, "Let's not work in the fields. We'll work in the mine. More money there" (37). Even they are exposed to their mothers' adultery. It is a blot on the Indian pious culture.

These are a few instances that show parents' indifference to their children's future. Further, it depicts the inclination of the villagers to temporary benefits. This makes a deep scar on the heart of Salvo Master, who continuously struggles to save the young generation from the evil of mining, which compels him to bundle his clothes and leave the village. Thereafter, no one is there to save villagers and their children from doom. The parents have a parochial attitude towards their children's education because they are very much engrossed in mining work. Being short sighted no one tries to stop Salvo Master from leaving the village.

2.3.3 The portrait of the Third World woman peasant

Another trope of ecofeminism is that women are saviours and close to nature because of their dependency on it and they are the first-hand victims of destruction of nature. To delineate this concept Naik, has sketched a major character Rukmini – the wife of Pandhari.

Rukmini represents the Third World women peasants' organic farming. The process of preparing seeds visualises the subsistence practice:

> Rukmini poured the paddy sprouts from the gunny bag on to the dung-smeared floor. She could feel the tiny shoots pricking her palms as she spread them out. She'd seen the first few sprouts that evening when she'd sprinkled the cloth bundle with water for the last time. The smell of the ragi grains being milled on the verandah mingled with the musky odour of the sprouts and an unmistakable sweet scent permeated every corner of their house.
>
> (10)

This same line of thought can be traced in Datar's perception, "This is especially true in a country like India where more than half the cultivable land is under subsistence production and is cultivated mostly by women" (Datar 110).

Rukmini is very much associated with her field. She is not ready to accept mining in place of their ancestral occupation of farming. The decision of Pandhari to join the mining disturbs Rukmini such an extent that she loses her temper on her children's continuous queries regarding sowing. She furiously responds, "The day after? Why? Because my husband has lit a lamp? Or because he's setting out to burn this house down?" (14). Rukmini's anxiety shows the impending danger that is going to engulf the entire village. Further, Naik would like to project that the mining has not only affected environment but also the age-old serene culture and sanctity of women. Rukmini, who is a simple, down-to-earth and family-oriented woman, always busy with domestic and field chores – shifts her attitude to crookedness, decisiveness and falls prey to Babuso's carnal desire. It has spoilt not only her character but the next generation's too. Slowly, their mother's adultery is imbibed and inculcated by Nanu and Kesar. Thus, mining has spoiled not only one generation but many generations to come. Rukmini says, "This man has already made two or three trips to our house and each time he comes when Pandhari isn't home. He'll destroy not only my home but all of Kolamba, she thought bitterly. She would have given him a piece of her mind that very evening but the presence of that stranger, the Gujarati gentleman, forced her to hold tongue" (26).

The second part of the novel opens with Kesar's preparation, which is the most thematic move to aggravate the impact of mining, to go for mining work. Mining leaves the villagers to battle with poverty, prostitution and addiction to alcohol. It is the consequence of Pandhari's greed and temptation for money, which hits his family hard. Everything is shattered in the fierce whirlwind of mining. He has lost his son Nanu, who is very much interested in studies. His father's insistence makes him quit his studies and work in the mines. Gradually, he slips from the hands of Pandhari and becomes a driver and falls prey to all the bad habits and involves in liquor and women. Very rarely, he turns up to the house and is not even bothered about his mother's illness or paying some money to the family. He has lost all human instincts.

Rukmini is attracted by Babuso's tempting gifts and the offer of mining work to make more and more money. But having lost her health, she becomes bed-ridden. She becomes a burden on her daughter Kesar, who is at a marriageable age. So, Kesar works very hard in the mines and she too loses her chastity. Mining has not only affected nature, people and their occupation but also depreciated their moral values. It is very difficult for Kesar to get a boy for marriage. Later, she gets pregnant and runs away. When Pandhari works on the fields, he is rich and happy, but when he joins mining, he loses his land. He is not even able to work on the mining, as he contracts tuberculosis. To survive, he starts the business of a tea vendor.

2.3.4 Deep reverence for Indian agrarian culture

The Upheaval (2002) has given a brief introduction to the Indian agrarian culture and their worship of nature, especially, focusing on the agriculture related festivals in the region of Goa. For Indians, agriculture is not merely a means to earn; they have a divine faith and reverence for agricultural activities, which highlights the essence of agrarian culture. Let us have a couple of instances of celebration of farming-related festivals. For example, worshipping the lake indicates that God or certain divine spirit lives in the river, and to appease the spirit they perform a worship called 'the Spirit of the Lake.' Next, the villagers celebrate a few festivals: '*Barras*' is an agriculture-related festival of fertility celebrated once in 12 months. '*Chavath*' is the fourth day of the bright half of the lunar month. '*Malni Punav*' is the celebration of the full-moon night in the month of Poush. '*Meerg*' is a festival that the sun is said to enter this constellation around early June by which sowing of the seeds starts. This is the period when the first showers begin and hectic activity in the fields is initiated. '*Shigmo*' is celebrated to mark the end of winter and the advent of spring.

Further, this concept is assessed through the action of Abu's divine reverence for the river and nature, thinking that God dwells in nature because it is His creation. Therefore, he used to say that they have to respect them. Whenever women approach the river, he bursts out on them, "Don't you know of that woman who washed dirty clothes at that spot during her monthly cycle. . . . she kept washing them month after month and remained childless" (16).

One more aspect of Indian agriculture is organic and nature-friendly farming. This is very vividly projected through the preparation of seedlings: "the paddy sprouts from the gunny bag on to dung-smeared floor" (10), after sowing they sprinkle "cow dung water on the freshly sown seeds" (10). Next threshing and harvesting, "The workers spread a layer of hay on the ground and covered it with a number of reed mats. . . . They carried sheaves from their circular stack to the reed matting and began to stamp them vigorously, rubbing the ears of paddy into the rough matting to loosen the grains from husk" (41). From preparing seeds to harvesting, every activity is organic and nature friendly. Even to maintain soil fertility the mixed planation is followed, "The cashew trees were laden with fruit that swung from every branch. Mango blossoms fell to the ground and baby jackfruit grew bigger and plumper on the trees with every passing day. Clusters of coconuts at the top of the palms looked like large pots strung together, promising a rich harvest in the coming weeks" (45).

Thus, Naik intends to portray that the agrarian occupation is passed from generation to generation by providing an ample background through the celebrations and involvement of people of Kolamba in rituals, festivals and farming, which is the trajectory of Indian agrarian culture. One stroke of modernisation and industrialisation in the form of mining swipes away these ages together deeply imbibed culture.

2.3.5 In-betweenness: modern projects and native roots

Naik, along with the basic ecofeminist elements, has added a new dimension to the ecofeminist discourse by projecting a few men characters' in-betweenness. They are neither able to accept modern projects from the bottom of their hearts by cutting down their roots from their nativity nor to reject them outright.

Pandhari regrets his wrong decision to join the mining project, but it is too late to rectify it. He is responsible for the fall of his family. Nanu, his son, is blown up on drink and addicted to all bad habits. His wife Rukmini is bed ridden. Further, his daughter Kesar is pregnant before marriage and leaves the house because she is not able to face people. Pandhari's helplessness and his family's pathetic condition are consequences of his working at the mine, "Nanu doesn't give us any money. He's his own master. He earns and spends. So what income are you talking about? My wife has been bed-ridden for five years now. I can't lift a weight above my waist level." In spite of earning they lost their happiness, "This money that we earn from mine. . . . it's like water poured on a heap of sand. No matter how much you pour, it vanishes in an instant" (84). This confession of Pandhari indicates how men eventually realise the repercussions of irrational decisions that lead to their destruction.

Here, *The Upheaval* (2002) trajectories the temporariness of mining as a profession. In spite of earning a lot, people are discontented. On the other hand, when they were working on their fields, they did not get much crop and earnings, but they were quite happy.

2.3.6 Shortage of resources: threaten the survival and livelihood

Further, Naik gives one more dimension to his perspective as he acquaints us with the Indian agrarian culture that not only peasants depend upon agriculture for their survival but also many other villagers depend on them. For example, a barber loses his forefathers' occupation of haircutting, "Then how did you give the priest his share this time?" Pandhari's response to this question is, "I get my hair cut at Naveli whenever I find time. You are old and unsteady so Nanu doesn't wait for you either, he goes to Volvoi or Naveli whenever he needs a haircut" (73). Though a barber has neither land nor work on the land, he gets paddy and other grain in reward of his profession of hair cutting (barter system). Not only barbers but also many more artisans, who depend on fields, lose their livelihood because of the introduction of mining.

2.3.7 *Rhetorical trope: the deterioration of land and river*

Next, we will study the impact of mining on environment through literary tropes. Naik elaborates the natural beauty of the village in the opening lines of the novel to show the images of rustic beauty. These lines suggest that once the village was full of flora and fauna. In Naik's words, "Kolamba village nestled in the curve of the river Mandovi as snugly as a water pot fits against a woman's hip" (15). The core objective of Naik is to narrate that the entire vicinity of the village is full of natural beauty and the villagers of Kolamba are agriculture oriented. But the mining mafia pollutes the land and the river Mandovi.

The village Kolamba gradually loses its natural scenic beauty. The fields become barren. The cow sheds are no longer full of cattle, and the few animals that remain there make no noise at all. Only the hum of machinery and the muffled roar of barges are left behind. In the fields, even grass won't grow any longer. And rivers are full of dust, stones and residue of mines:

> During the rains the lake fills with water and overflows, its banks and all this area where you are sitting now is waist deep in water. The road there, that goes up to the mine at Shenori . . . during the rains, water gushes down that road into the lake bringing with it dust and stones and ore and the whole lake turns red like the paint we use on our walls. This is the water that runs into the fields leaving behind a thick layer of mud . . . mud that destroys the earth so no amount of manure or fertilizer can have any effect. The fields around Surla village were destroyed in this way. Now it's our turn.
>
> (131)

Therefore, Naik tries to paint the stark and grim reality of the impact of mining not only on nature – especially on the Mandovi River and the fields – but also on animals. Women lose their chastity and virginity and are forced to throw themselves in the business of prostitution, which makes them contract many diseases. Furthermore, mining is a temporary occupation for the villagers. After that what is the fate of people of Kolamba? Like Kolamba many more villages and like the Mandovi River many more rivers are under the evil impact of mining. What will they do? Naik leaves the novel with an open-ended question – What next? The Ganges is choked with industrial pollution, domestic waste and carcasses (Mukherjee 2010:168).

2.3.8 *The uniqueness of* The Upheaval *as an ecohumanist narrative*

In the last section, through an elaborate discussion we explored the core elements of ecofeminism to examine the treatment of ecofeminism. In addition, the novel elucidates how man too is a saviour of nature and oppressed by modern developmental attitude. Kamala Markandaya in *Nectar in a Sieve* gives an account of the above-mentioned two concepts to illustrate how Indian environmental problems are human problems. The same line of thought is traced in Pundalik Naik's *The Upheaval* through Abu and Salvo Master. Along with these two characters, Naik discerns the role of men in the Indian agrarian culture, for example, Shanu and Shanker.

The first character to be evaluated from this point of view is a 60-year-old man, Abu, of the village Kolamba. Abu is disappointed with the villagers' action of leaving behind their age-old occupation of cultivating the land and joining the mining project. In every possible way, Abu wants to save the land from pollution and the Mandovi River from contamination. He plays a vital role in the novel because he is always found regretting and he contends that mining stands for man's destructive nature. Abu is the mouthpiece of Naik in capturing the immediate ecological problems that may crop up from mining. Naik gives a special role to Abu because through this character he gets an opportunity to present a didactic view to the reader. The basic intention behind introducing this character is to create awareness about the environmental catastrophe among the villagers. Therefore, we can consider him as an environmental preacher and a scientist. Abu stays near the Mandovi River all alone. Nobody takes care of him. Abu is always preoccupied with the agrarian culture, the cultivation of land, the education of village children and their future, and the contamination of the river, without having any selfish motive. Abu is very nature friendly. His day starts with brushing with a Neem stick; he stays in the hut; and he has a few earthen pots and only a few clothes to wear. In every possible way, Abu wants to save the Mandovi River from contamination, "Oh! Move away from the lake! Abu scolded a little boy squatting close to the water." Further, if he catches any man or woman near the river early in the morning, he says, "Don't you know that the Barmo, the Holy Spirit, lives by the lake? You must keep the place clean. If you squat there and dirty the area, worms will crawl all over your buttock" (Naik 2002:16).

Abu makes a fervent appeal to save Kolamba from the menacing disaster in the form of mining. He knows that dark clouds are hovering over Kolamba and the consequence of the villagers to join the mining for easy money will take a toll on their age-old occupation. Here, he narrates a few stories. Among them he focuses on Kolamba's story, the story of the village and how Kolamba is blessed with the river Mandovi:

> Let me tell you Kolamba's story . . . the river Mandovi flowed along looping and curving lazily, when she saw this pretty little village. Like a mother who gathers her child to her bosom the river changed course and flowed by this hamlet that is how Kolamba nestles like a waterpot against Mandovi's hip . . . we have no shortage of water here. One can come here any time of the year and slake his thirst.
>
> (67)

Abu is disappointed with the villagers' action of leaving behind their agrarian culture and traditional values. They are allured by the mining work. This reveals deep and complex emotions. He is upset and asks, "Would our people forsake their fields to go and break stones at the mine otherwise? Would our women go and lie with those wretched strangers even though their husbands are healthy and strong?" (68). Thus, Abu is a balanced character. He always thinks of the welfare of the village, children's education, rituals, festivals, the river and the fields.

Salvo Master is yet another very important character, like Abu. Without any selfish motive, he gives children not only formal education but also practical and spiritual knowledge, and many more things to lead a compatible life with all other living beings. Besides, he has an ardent desire to teach the next generation the impact of mining on them and nature. As he comes to know that Nanu is not going to continue his education because of mining, he visits Pandhari to convince him to send Nanu to school. Further, he predicts the consequences of joining the mining, "You did. And others like you. By forcing the new generation to buckle down under heavy loads. To carry piles of stones on their heads. Where is Nanu? Where do you take him to earn money every day? This is age when he should be studying" (75).

Here, we can explore one more dimension of Pandhari's character. He is an epitome of the entire peasant community of the village Kolamba. We can see how Pandhari's agricultural activities are imbued with his devotional involvement and consistent work on his land. Pandhari is always engaged in strengthening the embankment around the field, ploughing the field many times in such a way that not a clod of his field remains unturned by preparing his land for sowing. In this context, Pandhari's dedication and concern for his land is shown in Shanu's dialogue, "Maybe you should hoe it and breakup the clods by hand" (4).

The elements of spiritual ecofeminism such as 'immanence,' 'interconnection' and 'compassionate lifestyle' (Starhawk 2002) are projected through Pandhari. He feels very happy and satisfied with his work in the field. This embodies his oneness with nature, "His feet were sinking well into the soft loose earth, he noted with satisfaction. The land was well prepared, as in the years gone by" (4). Pandhari is the one who finishes all the activities related to the field and the one who sows first. He works very hard on the land for cultivation. Not for the sake of crop or harvest but for compassion, love and devotion to his land and occupation. His land is bequeathed to him from his forefathers. Pandhari is not only preparing the land for sowing days together but also even mentally getting ready for it. He is obsessed with it: "Pandhari hailed a couple of labourers who had come to the gadi for a drink and told them to come to his field the next day. He asked one of them to bring nivolo and told the other to bring a bundle of fifty plantain leaves" (8). Rukmini prepares seeds by tying them in a cloth bundle and by tugging it at the grass rope. Nanu invites his master for lunch on the occasion of the celebration of sowing. The excitement and the involvement of Nanu and Kesar in the sowing show the enthusiastic involvement of the entire family in farming activities. Even he says, "I'll go to the field to sprinkle the cow dung water on the freshly sown seeds. . . . see, I've kept these torn clothes aside to wear tomorrow" (10).

Here, the novel introduces the reader to two things: one, agriculture is a joint venture. Therefore, men and women involve themselves in farming. Two, the Indian traditional way of framing is eco-friendly, "Rukmini poured the paddy sprouts from gunny bag on to dung-smeared floor" (10). Babuso proposes Pandhari that he should join mining with his cart to carry stone chips from Naveli to the river Shenori. Pandhari's concern for animals is aptly reflected in his reply, "But the bullocks will have hard time hauling such a load" (12).

Next, Shanu, the peasant of Kolamba, owns a patch of land almost half the size of Pandhari's plot. That year he harvests a good crop compared to Pandhari's because he doesn't join mining. Instead, he works on his field. Pandhari joins mining and doesn't tend his field. He says, "Don't look so suspicious Shanu. . . . I'm really happy to see your yield. You've worked hard, now you reap the benefits. All this hasn't just fallen from the sky! Can you hope to catch fish without getting your loincloth wet?" (42). Shanu's suggestion to Pandhari – about how to preserve seeds – displays two things. First, it is an organic traditional Indian farming. Second, being a man, he has the knowledge of the traditional Indian agriculture as well as has empathy for nature.

The exploration of a few men characters' holistic and harmonious relationship with nature makes us rethink about Ynestra King's conceptual understanding of ecofeminism, which is grounded in these words, "We are loyal to future generations and to life and this planet itself. We have a deep and particular understanding of this both through our natures and our experience as women" (King 1938b:10). However, in the Indian context men and women have a deep understanding and knowledge of nature because nature is their means of survival.

Though many ecofeminists propound that women are the sole sufferers of the exploitation of nature, Naik projects that men are also subjugated by the modern developmental attitude. Pandhari, who continuously carries the loads of chips and the pollution of mining, contracts tuberculosis and is not able to work on mines for two reasons: first, he has lost his bullocks. In addition to this, the introduction of trucks in the place of bullock-carts, and big machines with shovels, conveyor belts and cranes and the jetting with moving belts that carry the ore right into the barge grabs the work of the workers in the mine. Moreover, he loses his ancestral occupation of farming.

Second, Nanu, who is very much interested in studies, because of his father's pressure he quits his studies and joins mining work. Gradually, he falls prey to all bad habits and gets caught up in liquor and women. As one day, Nanu carries a truckload of ore early in the morning. He parks the truck to unload it by the workers. Meantime, he wants to take rest but there are no trees in nearby provinces; everywhere one can see only huge mound of mineral dust. He wants to take shelter under the mound from the hot sun. As he is standing near the mound, suddenly the mound slides down on Nanu. He is buried under the sand and meets with his death.

Thus, Naik dissects the pragmatic environmental problems of India. Each page of the novel is tinged with the impact of mining on nature and human beings. Next, we study the same line of reasoning in Shivram Karanth's *Return to Earth* (2002).

2.4 Shivram Karanth's *Return to Earth*: the impact of modernisation on agrarian culture

Now, let us undertake an ecofeminist reading of Shivram Karanth's *Return to Earth* (2002). It was originally written in Kannada as *Marali Manige* (1941) and later translated into English by Padma Ramchandra Sharma. It depicts the impact of modernisation on agrarian culture. Shivram Karanth, a highly acclaimed Kannada writer, was a multidimensional personality. He was a *Yakshagana* artist, educationist, environmentalist and an anti-nuclear activist. He has written 400 works including of 45 novels, 231 tales for children, 4 short-story collections, 2 volumes of articles, 6 travelogues, 1 dictionary, 9 encyclopaedias, 2 autobiographies and 13 critical works on art and science (Karanth x). Karanth's novels recount a realistic picture of life with a critical perspective. They are imbued with multi-layered themes the most common of which are human relationships, irrational faith and belief, humanism, untouchability, irrational spiritual and religious activities and a concern for environment. Though he has written many novels, we shall briefly examine a few of his major novels that give a picture of his style of writing. These novels are a kind of introspection of life, society and environment. *Mookajjya Kanasugalu* (2012) interrogates moral values of life in the backdrop of the story of a widow and her dreams, *Chomana Dudi* (2010) focuses on Dalit issues, *Chigurida Kanasu* (2014) depicts themes like generation gap and issues related to ecology, *Bettada Jeeva* (2014) describes the pre-independent lifestyle of the tribal people in the forests of the Western Ghats and *Sarasammana Samadhi* (2015) advocates feminism.

Return to Earth (2002) is an archetypal narrative of the impact of colonial modernisation on the village Kodi. Especially, it focuses on the impact of English education on agrarian culture as a result of which most of the young generation from the village Kodi move to the city to get English education. The novel presents the story of three generations from 1850 to 1940. In the first generation, Rama Athiala – a priest and farmer – is more focused on worldly ambition, which makes him neglect his ancestral occupation. He sends his son Laxshminarayan (Lachcha) to the city for English education, which is considered a symbol of urbanised modernity. Along with this, the novel gives a vivid description of Parothi's (the wife of Rama Athiala) and Sarasothi's (the widowed sister of Rama) involvement in the domestic chores and fieldwork.

The second generation is represented by Lachcha, who is least bothered by his ancestral occupation and is lost in the trap of gambling, prostitution and deadly diseases, which forces his wife, Nagaveni, to abort her child. Further, the novel portrays Nagaveni's struggle to survive in the tough ordeals of life. The third generation is represented by Lachcha's son, Rama. He too is attracted by the glamorous urban world. He goes to Madras and later to Mumbai to avail education and a job. So, no young men are left at home to look after the lands. Only the women struggle and strive to work on lands.

2.4.1 Marked effect of colonial modernisation on peasant lives

Let us contextualise Shivram Karanth's *Return to Earth* (2002) within the framework of ecofeminism. Shivram Karanth is a critic of development and globalisation and speaks the language of the agrarian community. The critical analysis of the novel and the exploration of the characters in detail give a vivid picture of ecofeminist elements. The first key concept of ecofeminism – the patriarchal developmental attitude – is the colonial modernisation in the form of English education, which is portrayed through Rama Aithala, Lachcha, Rama, Sheena Mayya and Bava (brother-in-law of Rama Aithala). Rama Aithala, a priest, pays visits to nearby villages. He is very busy in his priestly duties from morning till evening and if he gets any leisure time, he would like to supervise what is happening in the town. He occasionally looks after the fields; but most of work in the field is taken care of by his wife Parothi and his widow sister Sarasothi.

Second, Aithala's family has a cow named Gopi, which always delivers a male calf. Parothi doesn't want to sell the fifth male calf. However, Rama's perspective on either to nurture or to sell the calves depends upon its economic benefits. Rama says, "Parothi, let us give off this gudda to Soorappa. Your Gopi will never give birth to a female calf. Can we keep looking after these male calves and drive carts? Let us give it to him for we can't let Soora go. He is the one who gives us the buffalo for ploughing every now and then" (14). And so, he sells off the calf. Here, the juxtaposed views are depicted.

Next, the modern developmental attitude endorsed in the novel is Rama Aithala's intention to send his son to English education in Kundapura or Udupi. This was an era when everyone longed to put their children in these schools with the hope to see them as a lawyer or a tax collector or in any other white-collar job with city job opportunities. Karanth remarks that during the period of independence English schools were started in Kundapura, Udupi and Mangalore (the Western coast of Karnataka state). English education becomes an epitome of modernisation that does irreversible damage to the agrarian culture of the village Kodi. Rama Aithala deliberates, "It would be good if Lachcha learnt some English. He shouldn't be stamping on mud and slush day and night" (138). Rama Aithala's obsession to send his son Lachcha for higher education has taken a heavy toll on two things. One, the boy is deprived of his fatherly profession of priesthood and connections with the soil. Second, this depreciates the pious Indian culture and lands him in all bad habits, which inculcate and imbibe treachery, deceiving and stealing habits.

On the other hand, Rama's desire to send his son to an English school is a counterpart to his wife Sathyabhame's craving to send her son to Shankara Aii's school to learn the salutation to Ganesh and read aloud the Bhagavatha or Ramayana. Rama is under the impression that "It would be good if our son becomes a lawyer. A tahsildar or an officer in the revenue department or any government servant – even that will be enough. Will the prestige of such a job ever be attracted to this priesthood?" (136). This central thematic element tracks the novel's exposition of the villagers' attraction to the glamorous modern world. They think that to work on the fields and to perform priestly duties are mean jobs compared to working in the hotels in the city because it gives more monetary benefits.

Though Rama Aithala sends his son to receive English education, he is not able to understand his son's changed temperament and the consequence of English education. He espouses his indignation in the words, "What sort of English is this? Can't he stay at home to offer a pot of water to those who come?" (160). Karanth illustrates that modern attitude has not only marred the agrarian culture but also the Indian culture.

Karanth tracks another character in this line, Lachcha. His personality is caricatured by his father's ambition of sending him to an English school. Lachcha develops a kind of indifference to his family and his village Kodi. Especially he is indifferent to his father, who spends his entire life to collect every penny to send him to an English school. He doesn't want to spend even his vacation in Kodi; instead, he wants to go to Padumunnooru (grandparents' town) or Udupi. The restlessness of Lachcha in Kodi relates his in-betweenness. He neither accepts the culture of English education nor synchronises with the village life: "I can't stay home without any work and it is getting difficult to work in the field as well. It is cultivation just for name's sake. Anyway, I shall go to Mandarthi because of our son-in-law's insistence" (209).

He falls prey to bad habits such as gambling and prostitution. However, he completes his degree and gets married to a lawyer's daughter from Mangalore, Nagaveni. He becomes bankrupt, "Starting a hotel in Mysore, Bangalore and wherever else he went, borrowing money, gambling, and then running away from there – this is what has been going on" (234). Lachcha's flamboyant quality has taken a very heavy toll on his family. Due to his disease, Nagaveni aborts her first child. Second, his father passes away in his worries, leaving the land in the name of Nagaveni. Third, he deceitfully transfers the land to his name and later sells it out. As a consequence, Nagaveni, along with her son Rama, has lost the means of livelihood. Leaving them in darkness, Lachcha elopes. All these are the consequences of whims and fancies imparted by English education.

The next character, the brother-in-law of Rama Aithala, who has a very different orientation about prosperous men, puts forth the basic contention of ecofeminist discourse. He thinks that a rich man does not have to go to field and work in the mud. He sees Rama Aithala, who works in mud: "He too had grown up among farmers, but not like this. They had given their fields on lease and grown up in comfort with they received from it. So he was slightly puzzled about his brother-in-law. He had heard that Aithala was a prosperous man. Getting immersed in the mud didn't appear proper to him" (82). Further, the carpenter, Hanumachari, is indifferent to the trees and always thinks of comforts. He, says, "Aithalare, you don't have a big box in your house to keep rice. The hall doesn't look good without it. Look, there is a jackfruit tree at the back. It has grown all twisted and is no good for making boards. If it is cut, a good box can be made" (90). In the same context, Sarasothi's idea of cutting down the tree juxtaposes Hanuma's view, "Don't listen to Hanuma's words and axe the tree. . . . The box is just a pretext" (90).

A few more characters reflect the ubiquitous prevalence of the patriarchal developmental attitude: Sheena Mayya and his son display a feeling of pride about their house in the words, "Appayya, the tiles on our housetop can be seen from Thekkate in the north, Hangarakatte in the south and Sastha in the east" (98). Sheena sends his sons to city to earn heaps of money so that he can build a big house and lead a comfortable life. These are the traces of ecofeminist contention of patriarchal developmental attitude.

Another feature the novel proposes is the application of globalisation in the setting of village that is the cold war between the global/local. Karanth's ecological sensibilities have found their way into the novel by showing these two concepts through the contrast between the natural and the artificial. The natural concept is shown through their houses: "Their house was on the edge of a sandy beach, and all around were cashew trees. The house had a thatched roof, not tiles; it was a wonder that it had survived the buffetting of winds from the sea" (1). On the other hand, the introduction of modernisation in the midst of the glory of nature is projected through the construction of tiled roofs by Sheena Mayya, Subraya Upadhyaya and Rama Aithala, which impeccably, resonates:

> But one longing he has in his heart didn't disappear at all. It has been there ever since Sheena Mayya's tiled house rose amidst their dwellings. But the number of those who had tiled houses after making money in Bangalore is increasing. If one surveyed the houses from Kodi to Manooru, at least half a dozen such houses could be seen. Subraya Upadhyaya had tiled his house as well. Rama Aithala was still bothered by his unfulfilled longing. Every time he talked about the village, the issue of tiled houses cropped up. (210)

Before Rama Aithala dies, he too fulfils his wish of building a storied tiled-roof house.

In the next section, let us trace another element of ecofeminism by scrutinising the women characters in the novel. The novel is a saga of the struggles and strivings of the Third World peasant women – Parothi, Sarasothi, Sathyabhame, Nagaveni, Bachchi and so on. The novel highlights that for these women to work on the field is not only a means of subsistence but also their close association and intimate bonding with nature is inseparable in spite of many ups and downs in their lives.

2.4.2 The saga of Third World women peasants

Karanth presents that women are conservationists through their various domestic and daily activities. These women spend most of their time either looking after the fields or taking care of the cattle. Paraothi and Sarasothi look after the cattle by preparing gruel by mixing hay and barn, milking the cow, driving out the cattle to the fields, gathering cow dung from the cowshed to throw into the manure pit. Next, for Paraothi and Sarasothi the most herculean task is to collect firewood. The collection of firewood is very much nature-friendly. They are not cutting down trees but they collect dried hone and cashew leaves that have been shed in their yard. These withered leaves are used to light the hearth to cook food. The remaining dry leaves are spread on the floor of the cowshed, which is later used as manure. Further, they collect firewood from the river as Karanth delineates:

> The river flowing from distant forests brought along with it a lot of firewood and trees floating along. It wasn't an easy job to go there, grab them and bring them back. To do so, it was necessary to leave the estuary and go further north for two miles along seashore. Notwithstanding the force of the rain, one had to leave home early in the morning and walk on the beach to be able to see what Varuna would bring. Sometimes, branches and boughs would arrive, and at other times, large beams. As soon as the wood was sighted, you had to jump into the water, bob up and down among the waves, and then grab it. If you waited awhile, it would come ashore by itself.
>
> (6–7)

The collection of firewood is a tough task; but in this work Sarasothi is on par with men by swimming in the river. She collects firewood along with her brother Rama. The collected branches are carried home and left out in the sun to dry. They have other jobs on hand, "pulling out manure from the field and raising the banks in their garden to the east of the house the day after the horsegram had been harvested" (8). Being women, they never hang back but are always ready to pick up the hoe and erect the thorny fence to protect crops. Once the bed to sow seedlings is ready, they sprinkle ash with some sand on the roots. Later, they water the plantation regularly. Before the completion of transferring the manure to the fields, they start the transportation of mud to the field; if it dries up, it becomes a very difficult task for them to do and they don't hire any paid labour to work on it. Rama Aithala, Paraothi and Sarasothi go together and plant the boat in the water and start scooping the mud into the boat. The process of carrying mud takes hours together. Parothi was involved in fieldwork before her marriage also; the only difference is that in her father's house she used to work with her father and here she does it with her sister-in-law. These two women are always bothered about cultivation: "Aithala did not carry the weight of household problems the way two women did. His life was on an even keel. Because of his priestly duties he paid little attention to farming" (12).

Now, let us focus on their renderings and humane treatment to animals. Parothi spends much of her leisure time communicating with the cow named Gopi, as if it is her kin. In this act, she portrays herself as natural element. She talks to Gopi loudly, "Such a fools these are! Gopi, the sun is shining and you should be grazing, like bathing where the bridge has given way; instead, you want to dance to a rhythm? What if it rains?" Perhaps, Gopi understood what she meant. She started grazing furiously. What a haste!" (19). Aithala considers this a "loss of her normal intelligence." He further adds, "When he saw her, he would say that she didn't need people to talk to. A wall would do as well. It wasn't mockery" (14). The women's reluctance to sell the calves of Gopamma, their cow, very clearly shows their concern to animals. Whether they are useful or not, they love them. Even those cows understand the language of humans. As soon as the cows hear their mistress Parothi call out to them "Enough eating! Kempi, Champi, come home" (20), they come running back. The mutual bond and reciprocal communion between women and animals show, substantially, their participation in the realm of the non-human world.

Karanth sketches one more woman character, Sathyabhame, the second wife of Rama Aithala, who is involved in the fieldwork. However, her character should be read through her desire to give her son, Lachcha, the very Indian education in place of English, which is influenced by the colonial modernisation. The women's opposition to the introduction of modernisation depicts the long-term result of this vicious attitude that may disrupt their lives.

The introduction of Nagaveni, the wife of Lachcha, in the novel is very much ecofeminist. The impact of the colonial modernisation – Lachcha's English education – especially, has taken a toll on Nagaveni's life. Lachcha – from a simple serene place Kodi – is sent to get English education to the city, where he is attracted to a glamorous world and gets infected with disease. The repercussion of this is Nagaveni's abortion and a disease that takes a long time to cure. Due to his education, Lachcha is not able to live in Kodi and goes to Madras, leaving her in Kodi and never heeds to her tearful pleas sent through letters.

Whenever he is short of money, he comes back, but he is not able to cope with the lifestyle in the village and so, he is not able to stay there for a long time. However, Nagaveni's rejection to stay comfortably in her father's house and her firm decision to remain in Kodi speak of her association with and divine faith in the land. But for her survival in certain inevitable circumstances, she leaves the animals in Kodi and goes to Mangalore. Yet, the bond between Nagaveni and the animals is not effaced. Here is a close association mapped in a description, "Kapila was lying down there in the shade. She got up when she saw Nagaveni. When they sat down, Kapila stood there locking her arm, tickling her. Seeing the affection showed by the cow, Nagaveni said, 'Bachchi, how much love and memory these dumb animals have!'" (315). For a while, life's very hard ups and downs make these women unconcerned about the land. But they are not able to retain their remorse for a long time. Sarasothi recounts, "Why this pain? Let those who have eaten, continue to eat. Until God calls, we shall spend our days chanting His name" (220).

Lachcha gets a government job in Bellari. Owing to his extravagance, he is neither able to send money to Nagaveni nor take her with him. However, "At home, she would be trouble-free, thanks to the income from land bequeathed by her father-in-law" (220). After Rama Aithala's death, these three women are left at home. Sarasothi and Nagaveni are able to work on the fields, whereas Sathybhame has not had much experience of working on the fields. Anyhow, Sarasothi and Nagaveni during the rainy season manage to do normal farming by planning for the future: "Nagaveni thought they should cultivate for three measures in future. Huruli should be planted that year and avare as well. It might not be possible to irrigate the cucumber patch. So, they could drop it. She knew nothing of agriculture. But she agreed with Sarasothi that sitting idle at home and eating was no good" (229).

Lachcha doesn't turn up in Kodi for three years. In the meantime, Nagaveni loses her second son. This is a second blow to Nagaveni to withstand. Nevertheless, she involves herself in cultivation, which is a healing poultice in her hard times. Further, we can trace how the women are strategic in their management of their cultivation, "It was decided that the older women would go on the journey. After sowing and transplanting were over, they would leave and be back in time for harvesting" (235). In spite of a lot of turmoil, obstacles and treachery of her husband to transfer the land to his name, Nagaveni tries hard to stay alive. But, one day she loses her patience and tries to attempt suicide, but ultimately she thinks, "We shall stay here whatever happens. We shall carry on working in the fields we have got on lease?" (260). This shows woman's divine faith in land. Nagaveni and Sarasothi with the help of Soora's family prepare three patches of land for cultivation. "Their fields turned green in three days. The only thing Nagaveni didn't like very much was the fact that Soora's family hadn't come over to collect their payment" (264). She is always involved in the cultivation of the leased land by watering, carrying ash, irrigating the field and growing vegetables by encouraging mixed cropping to retain the fertility of the soil. Simultaneously, they plant grains like sesame, avare and huruli.

Through hard work and a lot of faith in the land, Nagaveni is able to look after her son's education and is able to regain the sold land. "Nagaveni's joy was beyond description on the day the property was registered in their name again. Twenty years ago, Lachcha had told her that he would take her to Udupi but had taken her to Brahmavara instead and had had her property transferred to his own name. Within a month, the property had been lost to them" (429). Though the family of Soora takes other's land on lease, they are with Nagaveni in all her hard times without taking any payment for their work. Being poor they are rich. Bachchi – wife of Soora – says, "Amma, the debt we owe to land, to water, is there only if Brahma wills" (312). Next, we can view that a woman's reverence to land not just working and having concern for land but they are not willing to accept the modernisation in the place of their traditional occupation of agriculture. This can be tracked through Bachchi's perspective. She posits, "What? Leaving the town and coming here? When people here get fed up and lead towards the town? Should they come here wanting to eat palya made from avade greens? Have you seen our village now? As soon as boys go through Upanayana, they run to Bangalore. Now this is a place for girls and blind old people like me, or those who have no way out" (317). Further, her request to her son to return to the village shows her religious faith in the land: "Enough of your hotel. Let us cultivate the little land we have and live" (317). This explicates the ecofeminist stance and gives a picture of how colonial modernisation exploits agrarian culture including women.

Karanth shows that basically men are also equally associated with nature, but the evil in the form of English education deprives them from their ancestral occupation. Lachcha's cultivation brought in a lot of paddy. This shows his knowledge of agriculture:

> The rainy season stepped in slowly. Lachcha's agricultural efforts showed green in the fields. The same freshness of mind filled the household. Lachcha behaved in a free and friendly manner at home. . . . The crops weren't destroyed that year. Lachcha's cultivation brought in a lot of paddy. . . . Paddy was heaped up in the outer verandah. In the yard rose a haystack as tall as four men standing one on top of another.
>
> (247)

Here, Karanth's narration very clearly implies that if Lachcha had not gone to Kundapur and Udupi for English education, he might have led a peaceful and pure traditional life. Even Nagaveni might not have faced the horrible consequences. She might not have aborted and caught the deadly disease that took a lot of time to recover from. Placed in such agony and frustration:

> She was amazed by the touch of his hand. Ever since her husband came back, she had not been his companion at night. The very thought made her shake with fear. That night, who knows why, she could not put up a fight against him. Didn't she desire his company? Yes, she did. She would often think that she was a widow though her husband was alive. But she trembled thinking of the consequences of any such intimacy.
>
> (247)

2.4.3 The uniqueness of return to earth as an ecohumanist narrative

Karanth portrays the feministic concern in the novel through women's dependency, intimacy and faith in land. At the same time, it shows that man is also a nurturer of nature, which implies that Indian environmental problems are human problems and not just gender issues. The novel affirms the role of man as a nurturer of nature. Aithala's grandson Rama is another example who loves to work on the field in spite of earning a degree. He could have preferred to work in the city; however, his concern for agrarian culture and his contemplations on the importance of agriculture show the transmutation in Rama's attitude. Though he has English education, he realises the value of agrarian culture and gives up his craze for working in the city. He regains the land that is sold by his father because of his flamboyant nature: "Now they were overjoyed that they had regained their bond with the soil" (429). And with new energy and vigour, Rama starts working on the field with his mother, Nagaveni. "Amma, no need for vegetables this year. Let's grow some greens in the yards and be done with it" (429). Before the rainy season starts, he utilises the land on the beach for cultivation, but Nagaveni wonders how he can grow paddy in the sand. The response of Rama is "the land is mostly sand and there we can plant Gaali or cashew seedlings which may grow well" (430). Being highly educated, he also has a good traditional knowledge of the cultivation of land. Therefore, he asks Chenna, one who is helping him to work in the field, "Put hundred baskets of fish manure here, I want to grow tobacco here as well this year" (430).

Rama says, "He will carry on with cultivation this year but that he is not keen working on the field if the lease stipulates seven measures. He hasn't as yet signed on any paper for the lease this year. It is time for sowing. Shall I cultivate it myself? Is it possible for me to do it?" (373–374). Rama starts his work on the field: "They went down the beach and came to the newly made seedbed. There was a fenced tobacco backyard and a tank there. Rama and Nagaveni got into the water, filled the pots and started watering the tobacco plants" (420–421). Rama wishes to marry a girl who can look after the farm. He enquires, "Amma, does their daughter know how to water tobacco and harivé?" (435).

Soorappa, one who labours on Aithala's field, starts his day by working in the field by getting the paddy seedlings ready to reap the harvest. He does not work on Aithala's field to earn or support his livelihood due to his concern for land and his occupation, which has deteriorated owing to modern attitude. Thus, Karanth extrapolates man's impulse to serve nature. Indian agriculture needs the involvement of both man and woman: "Soora for ploughing, and his wife for transplanting. Sarasothi paid for debt owed to Soora's wife by working alongside her on her field so that it couldn't be said the field fell fallow because she worked for others for free"(28).

The craze for colonial modernisation and English education damages the agrarian culture. However, at the end, the novel depicts the return journey of Rama from the city to the village. We have to return to earth that is our base and root and this completes man's circular journey. Ultimately, he chooses the age-old agrarian culture, "Ayya, meshtre! Even after learning English, you are digging here in the sand. Why does my boy need any English? We, Mogera children, always speak only in English in the evening!" (430).

Thus, the ecofeminist reading of Karanth's *Return to Earth* registers the strivings and struggles of especially the Third World women peasants for survival and subsistence along with a few men characters' concern for their land.

2.5 Sarah Joseph's *Gift in Green*: a toxic discourse

Sarah Joseph's *Gift in Green* (2011) is one more novel in this line of thought that explicates the impact of construction of a mega city and a bridge on the paddy fields and water bodies. *Gift in Green* was originally written in Malayalam and later translated into English by Valson Thampu. Sarah Joseph, a

Malayalam writer, a feminist and a social/environmental activist, has written six novels and many short stories and essays. Her remarkable novels include a trilogy *Alahayuda Penmaklal* (1999), *Mattahi* (2003) and *Othappu* (*The Scent of the Other Side* 2005). The major themes of her novels are the dismal condition of the marginalised in the society and problems of women. She has won several awards too.

The plot of *Gift in Green* (2011) is not linear. It is a mosaic of many characters and events. No character is fully developed. All these are connected to water bodies and paddy fields of the village Aathi. The novel is double focused. On the one hand, it depicts the struggle of people of Aathi to save their paddy fields and water bodies from the introduction of modernisation in the form of construction of a megacity and a bridge introduced by Kumaran. On the other hand, the novel presents the fisherman's son Kumaran's ambitious project to construct a megacity and a bridge in Aathi, which he thinks are hallmarks of development. This dual war between local/global is punctuated with Joseph's unique technique of story telling through seven stories based on the Zen and Sufi stories, which may be piecemeal and an intervention in the narration of the novel. However, these stories play a vital role to create awareness about the bad impact of our selfish developmental motive on nature.

2.5.1 Backdrop to the novel

Before analysing *Gift in Green* (2011) from the ecofeminist point of view, let us have a glimpse of the personal ethos of Joseph, which is based on her deep understanding of environmental problems. There are a few American women's movements to save Love Canal and its groundwater sources from pollution (Mies & Shiva 1993). At par in India is Sarah Joseph, who functions as an activist through her fictional narratives. Her motive is not to save just one river but to save many such rivers that are being polluted due to the dumping of toxic waste and emission of radiation and, consequently, are taking a heavy toll on precious lives. Joseph, in her interview with the translator which appears in the appendix of the novel, says, "There is an Aathi, I believe, within every human being: the primeval water-bed of our residual compassion of our rootedness as sentient beings. There is a frozen, subliminal Aathi within you. If you would experience *Gift in Green* (2011) as a loving invitation to discover this hidden and stay inwardly nourished by it, I shall feel happily rewarded." Sarah Joseph's narration echoes ecofeminist philosophy and goes on the same line of Karen Warren's – an American ecofeminist – thinking:

> Adequate analysis and resolution of such environmental issues as deforestation, water pollution, farming and food production, and toxins and hazardous waste location must be integrally connected to an understanding of the plight and status of women, people of color, the poor, and children . . . [to help] one understand how mainstream environmental practices and policies often reflect, reinforce, or create practices and policies

that devalue, subvert, or make invisible the actual needs and contributions of women, people of color, the underclass, and children.

(Warren 1999:xiv–xv)

Gift in Green is a constructive reaction against Joseph's own anguish, and a protest against the exploitation of land and water, which are the basis of human beings' subsistence and sustenance. It is about a "murderous covetousness that despoils the future of the generations to come." The novel is an outcome of her reading of the most influential Japanese scientist and water researcher Dr Maasaru Emoto's *Hidden Messages in Water*. It projects the connection of water with the individual and his collective consciousness. Dr Maasaru Emoto also worked to combat nuclear radiation in Fukushima. Secondly, Joseph's environmental activism owes its origin in her study of a few of Kerala's pro-environmental movements such as 'Silent Valley,' 'Plachimada,' 'Anti- endosulfan,' 'Aathirappaly' and a protest against SEZ.

Thirdly, her work throws light on the pernicious problem created by the landfilling of hundreds of acres of water-beds and mangrove forests in Ernakulam district to build a cricket stadium. Fourthly, her personal interviews with fishermen – at Valanthakkadu in Ernakulam district in Kerala – adds depth to her understanding of the issue of landfilling. The fishermen were picking fish and mussels and cultivating Pokkali rice. In their interaction, the fishermen told her that they earned Rs.300 per day. Then, she asked them whether they could earn more. They replied, "Why should we?" They said that they meet their needs every day. They had no need to make more money. She ponders over this. She thinks that those who are caught in the rat race of development cannot understand the sanity and spirituality of these fishermen.

Subsequently, the most inspirational evidence is the US bombing on North Vietnam. In this bombing, a mother witnessed her nine-year-old son's death in her very presence, which numbed her. But later she said, "We got used to the bombings. Now we don't sit frozen in fear in the interregnum between bombs. We sow. We harvest. We survive!" This incident inspired Joseph to pose the question. "Is it rice or is it a bomb that decides the survival of our species? A world free from arms may be a utopian vision. But it is still possible to fill the world with trees, birds, butterflies, fish, vegetation, and to replenish the fertility of the soil and the purity of the water. We must work and pray to make such a world real" (Joseph 2011:Appendix).

Joseph is not against development; but she criticises our present attitude towards development. The present notion of development is to favour the powerful and the rich by collaborating with multinational companies and encouraging global culture in the place of local culture. She says that development should not be measured by statistical abstractions but by empowering most of us and, still better, by focusing on per capita happiness and per capita peace. Further, she adds, "The crime of exploiting nature, to the extent of jeopardizing its survival, is peculiar to our species. Other animals are incapable of this sacrilege. The idea of progress we embrace must be broad enough to accommodate the creator's justice that confers on every form of life the right to survive with dignity" (Appendix). These strategic perceptions are an outcome of her interactions with the first-hand victims of environmental crises and her pragmatic involvement with various movements and protests.

2.5.2 *The impact of a megacity and a bridge*

Keeping the elaborate personal environmental convictions and experiences as a raw material, Sarah Joseph sketches this novel on a wider canvas by portraying various characters and using various narrative techniques to reach the ultimate destination that is degradation of environment because of the developmental attitude. We begin our discussion with the first intervention of ecofeminism, i.e., patriarchal developmental attitude that is introduced in the form of the construction of a megacity and a bridge in the serene Aathi. It is the root cause of exploitation of women and nature. Joseph shows how transition takes place in the village from a simple, rustic and nature-oriented life to a modern and materialistic-oriented lifestyle by the construction of a bridge, a temple and apartments. This transition has taken a toll on the water bodies and the mangrove forest.

Kumaran introduces the evil of modernisation into the village. He is a tenth class pass and a son of a fisherman from Aathi. He perceives the futility of age-old occupation, farming, which yields just a few grains of paddy, some fish and a few oysters by which he will not be able to build a decent house and lead a comfortable life: "Or move from a reed mat to cot? Or dream of having decent clothes to wear? Or, even of using street lights rather than palm-leaf torch at night" (18–19). Through Kumaran's longing, Joseph tries to put together two ways of life, that is, the village life with its eco-friendly 'reed mat' and 'palm-leaf torch' and the city life represented by 'a cot,' 'decent clothes' and 'street lights,' which symbolise development. Kumaran's attitude and his inclination towards development is a representation of the core motive of the novel through which Joseph would like to pose a question to humanity – What is development? Is it in the true sense a development? This is a substratum of ecofeminist stance. Let us try to get an answer to this question by examining the novel.

Kumaran cultivates the Pokkali paddy fields – ploughs, waters and prepares seeds. Yet, he always feels a vacuum and purposelessness in working on the field and farming fish: "Staying here is waste of time. If we are to get any work at all, we have to get out of this place and go somewhere else" (19). There is always a fervent appeal in his heart to go to city to earn more money and find comfort. His father's problem is: if he goes to the city, who will look after the paddy fields and fingerlings? His response to his father's worry is, "All I should do, according to my father, is to stay here fondling the fingerlings. What else can you expect from the old-fashioned people? I have to build my life. What have the fingerlings to do with it?" (22). He always thinks that there is no future in Aathi. If he wants to have a bright future, he has to go to city. He is least bothered about the field and water bodies; his indifferent attitude and remorse towards nature especially water can be read in these words, "In pot, it resembles a pot. Trapped in a pond, it tamely takes on the shape of the pond. What is this water you're talking about? Does it have any identity? Will it ever be something in itself? The thought of it make me sick!" (21).

His negative perception towards his father's occupation and Aathi makes him sell his land and go to the city. He is the first person of Aathi who sells the land and leaves the village, abandoning his old father and mother to till other's land for their livelihood. He even leaves Kunjimathu, to whom he is engaged, whose virginity he has ruptured. Least bothered about her feelings, he goes to the city to gain the materialistic comforts that are considered as signs of development. Kumaran leaves the village at the time of the daybreak, when 'Darkness was everywhere' (22). Joseph's allegorical usage of 'Darkness,' which is an oft-quoted word, symbolises and signifies that once the village was snuggled cozily on the lap of the river and the forest but something wrong is going to take place in the near future. The patriarchal developmental attitude is going to disrupt the village in such a way that darkness (doom) will hover over the village and eventually, engulf the entire village.

After 36 years, Kumaran returns to the village as an allusion to modernisation in the form of the construction of roads, bridges and industry in Aathi which gradually pollute the water bodies and land to such an extent that birds and butterflies flee away. Slowly the mangrove forests start dying, the aquatic life is at the stake because of chemicals and residues of the industries that ooze in the paddy fields and water resources and radiation is emitted into the air. The simple and down-to-earth life of Aathi is transformed into the worst place by the destructive impact of the modern world.

The first step of destruction of nature is by the entry of Kumaran's army, "Kumaran's army marched into the water. Booted legs kicked and splashed in unison. The battered water rose, splattered sideways and crashed, flailing its head. Never before had Aathi witnessed such an exercise!" (39). Kumaran's army crushes the riverbed and oysters at which women start wailing and crying. They enter the forest by crushing the undergrowth so fiercely that birds start flying towards the sky in order to escape from the soldiers. This diabolic entry of Kumaran's army into Aathi demonstrates the vicious stampede of modernisation. The first project of Kumaran is to construct a bridge. For this, Kumaran acquires the land of Ganesha Subramaniyam. Once there was clear and pure water in the lagoon, rivers, wells, ponds, hills, sludge, marshes, paddy fields and mangrove forest filled with abundant trees. But, these water resources have become muddy because of the construction of the bridge by which fish and oysters have become scarce and birds have choked to death. The root cause of all this is Kumaran's intention to build a township in Aathi. The construction of the bridge has taken a toll on the mangrove trees and prawns karimeen. Kumaran wants to build a paradise in Aathi. He landfills and levels 400 acres of land, which is inclusive of Ganesh Subramaniyam's 300 acres of land. Villagers cultivate Ganesh Subramaniyam's land as collective farming on the basis of lease, which is a means of survival for the people of Aathi. Due to land filling, collective farming has come to an end. Further, Kumaran buys the land and property of Ambu, Prakashan, and also purchases land from the government.

To start his project of construction of a bridge and a megacity in Aathi, he has to convince the villagers: "I ask you the same question I asked thirty-six years ago. What do you hope to gain by staying planted in water three hundred and sixty-five days of the year? A little rice, fish, oyster. How long will that do for you?" (42). The answer to this question by the people of Aathi is that the island has been untouched by modern human hand, which infers rape on nature due to modernisation. The island has been bequeathed to them by their ancestors to toil in the paddy fields in sludge and marshes. Land, water and the forest are their treasure and resources of survival.

Second, for the construction of the bridge, JCBs, drillers, tipper lorries, truck and many more machineries enter Aathi's serene land by carrying massive quantities of earth, stones, bricks, sand and cement and so, life in Aathi begins to lose its serenity. It affects not only the lives of the villagers but also mangroves, prawns, fish, fingerlings, mussels and birds, which are giving away their lives. Third, the people of Aathi have not only lost hundred acres of land of Ganesh Subramaniyam but also are not able to cultivate the remaining land. As the outsiders, especially the big investors have started cultivating farms of prawn in an inorganic way that oozes the chemicals in the water bodies. Consequently, the fish, the mussel and many more aquatic lives are at stake. So, hunger prevails and spreads like wild fire in Aathi. Kumaran is the root cause of woes, worries and hunger of the people of Aathi by snatching their means of subsistence; that is, their lands. However, he instructs his workers that the work on the construction should be given to the people of Aathi to save them from hunger. This is a paradoxical determination to rescue the people of Aathi from hunger. He is very proud of his work and thinks that the people of Aathi may recognise all the good things he is doing for Aathi. "I am doing for Aathi, my Aathi. Who stands to gain when this forlorn place – mere water and forest – is transformed into a township cracking with industry? Oh my Aathi, my own native place!" (156). Further, he thinks that his multi-crore mega projects will definitely stand steadfast for the development and progress of Aathi.

Fourth, the major melancholy of the Aathi people is that Kumaran bought six months' city waste (garbage) – dangerous and health hazardous lethargic chemicals and toxic waste – to landfill and level the waterbodies and the fertile land in order to build a township and industries. And it unleashes killer epidemic diseases like typhoid, diarrhoea and fever. It takes a heavy toll of 19 children. The people of Aathi never forget witnessing their children's death: "Till the last days of their lives, the heart-rending memory of their children withering away, even as they watched helplessly or struggled in vain to keep them from the jaws of death, would haunt them" (276). Kumaran is indifferent to the sorrows of people and remarks, and he orders, "Don't stop the engine. How much by way of compensation will we have to pay to the fellow's next of kin? If the factory stops working for a minute, how much will be the loss? Between the two, which is more profitable for the company? Paying the compensation.

The engine is not turned off" (273). Kumaran's inhumane attitude is blatantly commercial and influenced by global culture. "Kumaran is hell-bent on setting up a mega city to monopolize the pleasures, the profits, the materials muscle and the prestige it proffers, even if it involves disembowelling the entrails of Aathi. Aren't naïve peacemakers like you laughably irrelevant to his scheme of things?" (278).

Fifth, Kumaran supports the concept of global agri-business that is from micro (Aathi) to macro (global) in order to get commercial profit-oriented agriculture in the place of indigenous farming. Vandana Shiva says, "Global agri-business is now attempting to take over food processing by making fresh, locally produced food appear 'backward,' and stale food clothed in aluminum and plastic appear 'modern'" (Shiva 2010:xx). Kumaran says that he will start industrial agriculture and grow large quantities of crop across the world.

He does not follow traditional skills and trades but learned new global trades. "Whatever I do is for the world as a whole, and not merely for the benefit of my household or my tiny homeland" (Joseph 2011:284). Kumaran has a global perception and thinks that the local people are not able to understand his project. It is beyond their estimation. Kumaran is an ambitious entrepreneur having a high goal with modern ambition of landing on the moon. He doesn't want to confine himself to paddy fields, fish and mangrove forests. On the other hand, the people of Aathi don't want to understand his global language. Here Dinakaran, as the mouthpiece of the people, asserts, "We people lead simple life. We farm and fish only to meet our daily needs. For that, we need our water and our land" (284). Joseph tries to convey the message that we need to be local and try to meet our daily needs, instead of being global entrepreneurs for our greed.

Let us try to contemplate the perception of Kumaran and the people of Aathi about Aathi. This is a war between the civilised and the uncivilised world. Civilised people think that development in the pursuit of physical changes in the village is in real terms modernisation, whereas the villagers think that the life they are leading is by itself development. Kumaran – a symbol of civilisation and modernisation – feels pity that he has done great things in so many places but has not done anything for his native place and regretfully says, "The ferry bank lies still in the same pitiable state it was when I left. Everywhere, the same, drabness, the same destitution, and even same dullness on every face, How awful!" (40). On the other hand, the people of Aathi are surprised at the expression of Kumaran, "What drabness? What destitution?" The people of Aathi fail to understand Kumaran's underlined meaning of 'drabness' and 'destitution' because with whatever lives they are leading, they are quite happy. There is an air of suspension, among the people of Aathi, about Kumaran's sudden outburst of love because when his mother and father are on the deathbed, he doesn't turn up and even his family has not once seen his father and mother. He has not returned to his native land for decades. What makes him pity the villagers? And his sudden outburst and deep concern for the villagers makes them ponder over his activities.

The patriarchal developmental attitude is shown through, along with Kumaran, a few minor characters such as Baaji's father, Prakashan, Chandran, Komban Joy, Ambu and so on. Let us have a cursory glance at these characters. As Baaji's father has no resentment towards the project of Kumaran even he supports them in their endeavours in every possible way. He sells fish to the tent people (Kumaran's troop) to get a good price to his fish. When Dinakaran asks Baaji's father why he has to sell his fish to the tent people, just for a little monetary benefit, he responds, "it is a matter of my livelihood" (5).

Next, Prakashan, Ambu, Chandran and many other young men of Aathi are captivated by the bright plan of Kumaran's project, because they are fed up with water-life and want to be like Kumaran by being a part of Kumaran's project of building a township. Komban Joy, a treacherous man, dances to the tunes of Kumaran who asks him to borrow the land of Kunjimathu on lease. Kunjimathu's land plays a very vital role in the project of Kumaran. Kumaran bought Ganesh Subramanyam's 300 acres of land and the government land which was sandwiched by Kunjimathu's five acres Pokkali paddy field and 2 acres of forest land. Therefore, Kumaran wants to buy the land of Kunjimathu. On the instruction of Kumaran, Komban Joy, a dealer in plot selling, wants to lease the paddy fields of Kunjimathu to farm prawns. While farming, he uses endosulfan and other chemicals by which fish are choked to death. He is very greedy and indifferent to nature, "The moment Komban Joy admitted that it was endosulfan that he had mixed in the farm a few days before the kaappukalakku was to begin, in order to grab the whole of the fish harvest" (178).

He joins Kumaran's plan and forges the documents – title deeds of property – of not only Kunjimathu but several others too.

Further, Joseph predicts the transition of Aathi into a modern city through an analogy of the act of a magician. The magician is one among Kumaran's men. He swirls the magic wand in the air three times and in a fraction of a second everything vanishes. The bank, the ferryboats, the water, the paddy fields, the canal, the pond, the wells, the palm groves and mangroves are suddenly replaced with huge buildings, broad roads, hotels, cinema halls, parks and shining shopping malls. This is a snapshot of the sinister transformation of Aathi.

2.5.3 Within and without: women's concern for nature

The stance of ecofeminists that women are saviours, nurturers and caretakers of nature is clearly reflected through a few women characters such as Kumaran's mother, Kunjimathu, Shailaja, Karthiayani, Devaki, Divya, Gitanjali and others. Kumaran's mother is very much disgusted with Kumaran's indifferent attitude to his age-old agrarian occupation and Aathi. Her fear of her son's attraction to the city, modern life and his indifference to land and water are painted in her words of caution, "Don't forget your root, boy" (19). Kumaran's mother is always busy in weaving mats, looking after the cattle, cleaning fish, chopping fronds and so on. The next important character is Kunjimathu, the daughter of a fisherman. She too engages herself in cleaning fish, farming paddy and fish, digging basins around coconut trees, bathing buffalos, sowing bean seeds, working in the paddy field and farming prawn. Her world is limited to rice, fish, water, paddy fields and lake marshes. She works on her field very religiously. Especially, she had a lot of faith in water, as she says, "water knows everything and forgets nothing" (21).

First, Kunjimathu's thought process gives us the picture that the women's concern to nature is broad and foresighted and has basic human values and is not touched by the modern and materialistic world. When Kumaran wants to go to the city for his livelihood, she is puzzled. "What did Kumaran lack here? Even if you fished only till noon, you could make enough to feed a whole family. Fish or mussels: enough to meet one's daily needs. Add to this the two Pokkali harvests from the paddy fields every year that anyone could reap. It was quite sufficient to live free from want" (20–21). Apart from growing paddy, they grow very good quality rice, pumpkin, white gourd, spinach and beans, using Indian multi-cropping patterns. Further, they rear a cow, a buffalo, hens and ducks. Kunjimathu is not able to get any answer to the question on Kumaran's indifferent attitude to Aathi and his passion for the city.

Second, Kunjimathu's concern for land is shown through her action of selling her gold to get back the land of Kumaran, sold by Kumaran to Muhammad Kannu of Beypore, to his father and mother. Kunjimathu's life is her field. As Kumaran has left her and gone to the city for earning, she doesn't marry but dedicates her entire life to her field and water. She spends her life for farming and to take care of water-bodies, as 'immanence,' 'interconnection' and 'compassionate lifestyle' (Starhawk 2002) which shows the spiritual ecofeminists' essence:

> Kunjimathu, at sweet seventeen, stood enjoying the fragrance of the paddy field, ripe and ready for the harvest. The sweet aroma of her body caressed the water tenderly. Tiny fingerlings dated and danced all around her. The birds that came to peck the grain perched themselves on her shoulders. The sight of Kumaran terrified Kunjimathu, the birds and the fingerlings alike.
>
> (Joseph 2011:170)

Schematically speaking, this novel lays a foundation to the critique of the gradual introduction of corporate agriculture in place of organic agriculture. Kunjimathu is angry and disappointed with Komban Joy's new way of artificial farming – with the usage of endosulfan. Komban Joy's new way of farming not only poisons Kunjamathu's five-acre Pokkali field but also the fish of Aathi, which are choked to death.

Third, when Komban puts forward the proposal of borrowing Kunjimathu's paddy field on lease for prawn farming, she pours out her remorse. "No one farms prawn in this place" (174). She explains that with tides, thousands of little prawns come into the mangrove forest in which natural prawns as well as fish that get into the farm stay and grow there. Therefore, there is no special need to look after them. Joseph shows that the women are not greedy to get more money or harvest by following the new way of farming followed by Komban Joy. Komban Joy wants to use inorganic farming by spraying DDT to get more yields. However, the result is that thousands of butterflies and green frogs are killed. But, Kunjimathu wants to save butterflies, frogs, prawns and fish by supporting organic farming. Therefore, she gives a very strict warning and asks him to do organic farming by informing him pesticides and chemicals are not acceptable.

Here, we will give a cursory glance to two more characters, namely Karthiayan and Devaki. Both of them warn Komban that Kunjimathu's field yields fish and rice. He should stick to this and not use any new way of farming or use inorganic chemicals such as quicklime, potion, pesticides and fertilizers. Further, they have deep rooted religious faith and reverence for their land; for instance, to retain the fertility of the land they are ready to keep the land uncultivated, but at any cost they don't want to give it for inorganic farming. In practical terms, they are the real scientists. Next, Kunjimathu wants to protest to save water-bodies of Aathi, each of which is a treasure and source of livelihood for the people. Nonetheless, it is at stake because of the modern juggernaut. Therefore, Kunjimathu sits in waist-deep water as a protest, to save Aathi's natural life. She wants to sacrifice her life to save water-bodies, which are filled with garbage because of the construction of the township. She thinks that they are losing their connection to the water-bodies:

> Paddy fields, parched. Trees, dry and withered. The earth, cracked. Wells, dried up. Cattle, tormented by thirst. Birds, perishing. Children, howling in hunger . . . Desert storms raging with a vengeance. The burning sand it brought along, covering the land, Red-hot rocks. Thorny bushes. Scorching heat. Freezing cold. A woman who wandered about, aflame within and without in the wilderness: Kunjimathu!
>
> (196–197)

Though the project of Kumaran is providing them work and comfortable flats, the people of Aathi are not willing to go to new flats because they don't want to lose their intimacy with fields and water-bodies. Kunjimathu says, "I shall live working on my land and water. I'd rather die here, right now, than accept anything less. If all these water beds are landfilled and erased how can we work on the land and survive?" (203–204)

Let us examine another woman character, Shailaja. Shailaja is a nurse in the hospital. Shailaja suspects and questions Mary, the sweeper, where she dumps the leftovers of deliveries like placenta, umbilical cords, sanitary napkins, blood-soaked rags and cotton pads. Mary says that she buries them in the earth. But, the attendant reveals the truth: Mary dumps everything into a large pipe, which runs back to the bottom of the primal water. It disturbs Shailaja and starts enquiring about the root of the problem. She later comes to know that all the rubbish is getting dumped in the water: "They were not buried in the earth. Nor did they decompose to become manure for trees" (74). This displays her passionate desire to save water-bodies.

Next, let us have a glance at Shailaja's response to her husband Chandramohan's village Chakkam Kandam. After getting married to Chandramohan, she goes to Chakkam Kandam, which is poles apart from Aathi in its serene atmosphere and purity of water. Chakkam Kandam's waters – the wells, the ponds, the channels, the stream and backwaters – are all covered with layers of shit. Earlier, the Chakkam Kandam water was clean and pure, as Shailaja's mother-in-law narrates, "These backwaters were so beautiful! We could not take our eyes off them. Paddy fields that yielded harvests of gold, the lagoon waiting to be milked of bounties worth millions, fishermen and mussel pickets, boats loaded with coir, huge country boats going to Kodungalloor. . . . But later go on living with this shit-saturated water and the stink swirling in her nostrils!" (83 & 85). Shailaja is not able to resist the impurity and foul smell of Chakkam Kandam. However, for Shailaja there is no other choice left to deter the foul smell and impurity of Aathi water, "Plastic carry bags lay scattered like bloated fetuses. Over them, flies and mosquitoes droned with a vengeance" (207). She compares the present polluted Aathi with the old village, which was clean and bloomed with crystal clear ponds and lotus flowers. Farmers had water bodies measuring four or five acres of crop and she recalls women's involvement in the fields: "For generations, the people of Aathi, especially women had been planting paddy seedlings, and reaping and threshing the harvests. Even as the harvest season finished, they would begin their work in the fish farms. Shailaja, too, had worked in those paddy fields" (265).

Starhawk, the spiritual ecofeminist, propounds that the more we understand that we are a part of nature, the more we will understand our oneness with all that exists: human beings, natural cycles and processes, animals, and plants (Tong 2009:253). Shailaja's fervent appeal or religious feeling for nature is just a representation of the essence of spiritual ecofeminism:

> Wouldn't I have circled the land, cleaning it as I flowed, nourishing the grass and plants? From the earth to the sky, from the sky to the earth . . . seeping through the soil to the depths of the sky to the earth . . . creeping up through the roots to the tips of leaves . . . into you as you cup me in your palm and gulp me down . . . and back again from you into the earth.
>
> (Joseph 2011:208–209)

The people of Aathi gather to protest Kumaran's project of building a township in the village. In spite of the blood gushing out from their injured heads, the protestors block the road and remain firm under the lathi charge of the policemen. They cry, "Kill us. Kill all of us and bury us. We don't want your mansions. Our earth and our water – that's all we need" (217). Watching the policemen cruelly beating up the people of Aathi, Shailaja empties the kerosene can on her head and challenges the policemen that if they touch any one of the protesters, she will put herself on fire.

Another passing reference is Gitanjali, a woman from a city who comes to Aathi to cure her daughter's strange disease of the unfeeling metropolis. She sets out from the city in search of water bodies that could act as a healing poultice to her daughter: "When I set out, I had no idea I would reach a place like this. . . . It is Aathi that restored Kayal to me. I now know that water is the synonym for peace. Peace is healing and healing peace" (328). Further, Gitanjali discusses about the indispensable relationship between human beings and nature, "My Kayal is everywhere here. She is at one with Aathi. She and Aathi are one" (329).

Due to landfilling and levelling, most of the water-bodies and paddy and fish farms are beyond the reach of people of Aathi. But, this poses a challenge to them to come together and face the crucial and hard time to save nature. Therefore, all the women of Aathi come to a strong determination to do farming. Without intimating the men of the village, they start their work of cultivation. They collect mud from the bottom of backwaters and bring it to the fields – first a layer of clay, on top of that a layer of grass, and then another layer of clay. All these tasks men used to do but this time women struggle hard to build the bund: "On this day, however, the women worked on their own despite, their dearth of experience and the diffidence that nagged at them" (335). Vandana Shiva's assertion is in a similar vein:

> Women were the world's original food producers, and continue to be central to food production systems in the global South, in terms of the work they do in the food chain. Agriculture has been evolved by women. Most farmers in the world are women, and most girls are future farmers; they learn the skills and knowledge of farming in fields and farms. What is grown on farms determines whose livelihood are secured, what is eaten, how much is eaten, and by whom it is eaten.
>
> (Shiva 2010:ix)

Kunjimathu is a supporter to boost the morale of women. They finish building of the bund. They sow the seeds and later these seeds germinate on the mounds. Kunjimathu implores the women of Aathi: "Even if Kumaran offers lakhs of rupees and all of his gold, not a soul, not even a child, should leave Aathi" . This is not only Kunjimathu's deep love and devotion to Aathi but all the women retort the same thing in response: "Do you think we are happy to leave Aathi? Or that we'd give up our land if we had a choice? Don't leave Aathi. No matter what. You leave Aathi, you die" (Joseph 2011:338).

These are a few sketches that show that women have indispensable love, reverence, dedication to the water-bodies and land. *Gift in Green* (2011) paints the picture of impact of developmental attitude on land and water, which is the base of human subsistence and sustenance. Along with these core concerns of ecofeminism, we explore a few more aspects of ecological crisis.

2.5.4 Rhetorical narrative: the language of nature

In this section, the research endeavours to project Joseph's deep engagement with natural phenomena by narrating continuous metamorphic transitions in the water bodies of Aathi by man's inhuman attitude. Water is an ominous character in the novel. Joseph narrates the story as if the readers were rowing on the river of Aathi in order to understand the jigsaw in the lives of the people of Aathi. The metaphorical role Joseph gives to water pervades every page in the novel which displays the past purity and serenity of the water of Aathi. Joseph wonders, "It encircled Aathi, an enchanting world in itself, its waters cool and serene. Sitting in that rare world of impregnable silence" (25). On the other hand, Kumaran, the archetype of modernisation, erects a granite bund that goes against the natural current: "the water, wailing, howling . . . knocking its head against the granite bund" (203). It is polluted by polythene bags, liquor, cola bottles, leftover food, rotten fruits, matchboxes, mélange of decaying garbage and chemical residues to such an extent that the most of aquatic animals and plants are choked to death. When Kumaran starts to build the granite embankment and fills it with earth, then, the low-lying land of the west is submerged and water reaches almost near the threshold of the houses posing the question: "May I come in?" (206). All through the novel water is personified and it plays a vital role in depicting the environmental crises and ecological imbalance.

Second, the development of characters throughout the narration of the novel does not follow the conventional technique of writing a novel. In her interview with the translator, Joseph extrapolates by saying, "If I had accentuated the portrayal of characters any more than I have, they would have stuck out like sore thumbs on the even landscape of nature." This fictional technique exhibits her basic purpose of a unique way of narration. It is to emphasise on a thematic compulsion rather than focus on the usual development of characters. All her characters are figurative and symbolic in nature. The prophetic Noor Mohammed's fervent desire to save Aathi and its natural beauty is very clearly depicted through his stories. Markose, who is very sensitive to nature, projects his feeling through poems and tales that depicts the culture and psyche. On the other hand, Kunjimathu's and Shylaja's potent concern for nature and their contribution to nature are shown through their actions to save the river and the paddy fields. All characters are developed to spotlight deterioration of nature because of the modern juggernaut.

Third, Joseph uses a unique technique of story telling through seven stories, which may be piecemeal and an intervention in the narration of the story of the novel, to create awareness about our selfish motive and exploitation of nature. Each story drives home a message that relates very much to our indifferent attitude to nature. There are two purposes to narrate rhythmic and unpunctuated stories. First, to show the relevance of the story to the present by posing a question at the end of each story: how can we apply the story to present context? Second, all the stories are borrowed from the *Puranas*, the *Bible*, the *Quran*, the Zen and the Sufi traditions, folk narratives, historical events and from the life of St. Francis Assisi. Here, she scrutinises the role of traditional stories in the contemporary context. Third, she examines how all of our spiritual scriptures preach the oneness of all living beings, which we have left far behind. Many Zen and Sufi stories, especially, centred on water, provide a bottom line to the novel.

2.5.5 Aathi: the collateral victims of juggernaut of progress

Gift in Green (2011) is a deep abiding story of the mute Aathi, which forms the backdrop of the novel. By and large, Aathi plays the role of the protagonist in order to narrate the pathetic condition of birds, animals, nature and even human beings. Joseph portrays the picture of past and present of Aathi by comparing and contrasting the lifestyle of the people of Aathi. Once, the village had a clear and clean lagoon bountiful of fish, mussel and water plants along with a mangrove forest, which was full of birds, animals, and trees. But due to man's unacceptable actions it is rendered dirty, sticky and gluey. Aathi was rich especially with innumerable water sources such as wells, canals, water springs, ponds, water drains in paddy fields and slushy marshes. Joseph represents very beautifully the orientation of nature through the sweet sounds of nature and their interaction with water bodies which projects the harmonious relationship among all the natural things such as a falling of sallow leaves on water, blossoming of flowers, dance of the moss and worms emerging from their hideous places. One can inhale or even hear every minute detail of the bliss of nature and get enchanted. Joseph calls the mangrove forest that encircles Aathi village 'Greenbangle':

> The crabs in the mangrove forest were green in colour. Crabs of that colour were not to be seen anywhere else in Aathi. Not only were the crabs, the frogs, the butterflies, the grasshoppers and the snakes green in colour but even the wind that blew in the forest seemed green. For that reason, the forest was affectionately called the 'Greenforest.' Since the forest virtually encircled Aathi, it was also called the 'Greenbangle.' Eventually it came to be known as the Greenbangle.
>
> (210)

The rich allusions to animals, birds, flies, trees, even air and everything which is green shows that the village has bountiful natural resources which are enough for the survival of the villagers.

2.5.6 Deep reverence for Indian agrarian culture

The novel projects human beings' deep reverence to Indian agrarian culture. Dinakaran has deep faith in land and water-bodies, "We want to stay here and shield the waterbodies" (220). He has a great reverence to earth and water, "The earth and the water are our granaries, not Kumaran . . . His charity can never be a substitute for the land and the water he has stolen from us" (244). Dinakaran realises the importance of natural resources, which cannot be compensated or replaced by modern materialistic comforts. Dinakaran's attitude towards Aathi is not only as a means of survival but also as something divine and valuable, "For us, Aathi is not a pageant of fleeting sensations or a mere means of survival. It is an invaluable heritage, an incomparable experience . . . something that deserves to be treasured for all time to come" (292). He always thinks that the simple traditional ways of cultivation of land should pass to the next generation. The fields give livelihood to human beings as well as to birds and animals, "The sparrows will still need to find their subsistence. The parrots must be free to peck their daily ration of ripe paddy from the stalks." Dinakaran thinks that this is the privileged way of living of Aathi people.

Along with Dinakaran, Markose (the story teller), Ponmani (Dinakaran's friend) and many more block the way of lorries and tippers which are going to landfill the waterbodies. They want to save water bodies and paddy fields from destruction at any cost. Markose, a nature lover and a saviour, says to Prakashan who is the leader of project of landfilling and constructing a new city under the guidance of Kumaran, "You are cutting the branch on which you have to sit" (215). The raged Ponmani asserts, "Until our last breath departs from our bodies we shall not leave" (220). Ponmani desires to give voice to the voiceless – people of Aathi and non-human beings – to give them equal rights, "How can we fail Aathi in the face of such danger? I speak for all of Aathi, including the trees, the birds and the innocent people who know not how to give voice to their worries" (153).

Through the unique Indian culture of story telling, Joseph acquaints us with our harmonious relationship with nature in the past. Moreover, the painful danger is impending over the Aathi's serene land and there is transition from our eco-friendly relationship to the greedy selfish motives. Here, it is very apt to ponder over Mikhial Epstein's observation on human values, 'hypercritical and hypertextual' resulting in what he calls 'intellectual autism,' while earlier humanities served social ends (Selvamony 2014:7). Noor Muhammad, a story teller, narrates the stories to the people of Aathi. All the stories sprout from "the Pokkali paddy fields, rooted in the primeval soil and nourished by the water and the warmth of life. Sailing from place to place, narrating stories here, there and everywhere, they arrived like birds of passage, their feet soiled with the red earth, their clothes reddened too, and their hair dusty" (Joseph 2011:15). We can see man's deep rooted reverence and knowledge of agrarian culture by espousing the past experiences and their applicability to the present, "This is the story of the first evening. . . . What we have to do now is to consider how we shall apply this story to our lives" (17).

Joseph introduces two concepts – need and greed. If we focus on need, then no need of any development and modernisation. We can manage with our existing resources and means. In *Gift in Green* (2011), Dinakaran points out, "We were leading a peaceful life, harming no one, nor stealing from anybody's plate. All we had was what we had earned by the sweat of our brow" (204). But, when this need transits into desire and further greed it paves a way to cheat, hurt, exploit and kill, "Greed will make us thieves who steal. . . . We will cheat the land, cheat the water, cheat the sun and the moon. Cheat even the women we marry and the children we beget" (205).

2.5.7 *Shortage of resources: threat to the survival and livelihood*

Dinakaran is disgusted and disappointed with Kumaran's project of landfilling, especially the waterbodies, and burning of the forest and hill. Kumaran represents modernisation, without having any foresight, which just focuses on momentary benefits and comforts that are considered development. Dinakaran says that we need water and forest for subsistence and if man goes on plundering nature like this, what will be the consequences? He says, "what will happen if the forest is burned. You also know that we will all perish if the water and the marsh are buried under the earth you are out to dump on them. To betray Aathi – betraying your own mother could not be worse – you have gathered a suckling band of unsuspecting children who don't know their right hand from their left" (215).

The people of Aathi have been living happily on available natural resources. Kumaran's project of the construction of the megacity and the bridge at the cost of birds, fields, waterbodies and people of Aathi shows a juggernaut of progress. Kumaran and Ponmani belong to Aathi. But their attitudes are poles apart. Ponmani doesn't support development and the project of modernisation: "What is the need for this bridge here? A bridge only for private use, an exclusive access to the paradise that Kumaran intends to erect. . . . The building of the bridge should not be allowed to continue" (152–153). Kunjan Karnavar, an elderly villager, is very upset when he comes to know that the people of Aathi would sell their land, hoping to get a comfortable life with a good alternative for their livelihood. He calls an assembly and apprehends the importance of land. Further, he gives a cautionary warning to the people. If they sell the land to Kumaran's project, they may suffer shortage of resources: "Have you lost your senses? Alternative! Other than toiling on the land and in the water, Aathi does not wish to know any other alternative. You've chosen to sell and squander only because you no longer wish to live off the sweat of your brow. Go, all those whose hearts do not ache at abandoning Thampuran's soil . . . go" (288). Ponmani extends the same line of thought. He says that after selling their land they may get work as per their capacity. They may be allotted the work as gardeners, gatekeepers, or to wash cars and dogs of the landowners, to clean bedrooms and women have to work in kitchen or else entertain men.

2.5.8 *The uniqueness of Gift in Green as an ecohumanist exposition*

We have explored the basic tenets of ecofeminism in the novel. However, Joseph tries to show that both women and men are nurturers and subjugated by the development attitude. Sarah Joseph's narration demonstrates that man too is a saviour of nature. Dinakaran's concern for animals and birds is portrayed through his anger at his brother Siddhu and his friend Ponmani. Siddhu, with the help of his toy gun, makes a sound by which all waterfowl scramble up in fright. Dinakaran remarks, "Aren't you ashamed to scare them like this?" (6). Further, Dinakaran has taken action on birds trafficking; he suspects that something wrong is going on so he stops a car on the bridge and asks the driver to open the trunk. In the trunk, there are several jungle fowls "some of them still bleeding from their wounds. Many were convulsing with pain. Some struggled desperately to escape" (165). He regrets on the action of his own people who are showing the nests of the birds where the birds are hatching young ones: "It is our own people, lured by money. . . . That is the extent to which we have degenerated" (166). Dinakaran and Baaji are stunned by the massacre of water birds. They have a strange fear: "the dead water birds lying all around him. Neither bullet marks nor any trace of blood could be seen on their bodies. How, then, did they die? On the water, too, many dead birds floated" (167). It is a puzzle for them. How did these birds, fish and prawns die?

Later, they come to know that Komban Joy, who leases the farm, uses artificial pesticides by which many birds died. Dinakaran thinks that in the ancient times, elderly people toiled very hard from dawn to dusk for the sake of the next generation. "This paddy field is as soft and yielding as it is only because the sweat from our very bones has rained upon it" (163). However, these people are very indifferent to land and animals. Dinakaran, in every possible way, tries to save water bodies and paddy fields from the project of the construction of the megacity and the bridge.

In this novel, we can trace that man is also subjugated by the development attitude. Dinakaran, in his entire life, struggles very hard to save Aathi's paddy fields, the hill, the forest and the waterbodies from the landfilling and the construction of the mega city and the bridge, but everything is in vain. He sacrifices his life for the sake of the waterbodies and paddy fields of his village. In the end, his body, wane and weak and wrapped in a mat with every inch of it bruised, bobs on the waves. The people of Aathi fetch him out of water and give him water to drink. There ends his life. Here, Joseph narrates very intensely how man and nature are one and all. "Forest for forest, Fish for fish, Bird for bird, Longed and yearned with him to die" (347). An indispensable doom is waiting to engulf the village, "Silence primeval . . . like the placental rupture of darkness. The whisper of seeds sprouting in the dark, the aroma of clay from the Pokkali fields, and fingerlings playing in the waters of Aathi" (348). Dinakaran struggles hard to save Aathi from the havoc of modernisation but he becomes a victim of politics and loses his life in the modern juggernaut.

2.6 Conclusion

The aim of the present chapter was to scrutinise Indian novelists' treatment of ecofeminism and their unique contributions to the ecofeminist discourse. The aforesaid discussion of these four novels espouses three things: First, the intervention of tannery, mining, colonial modernisation and the construction of megacity have not only disrupted pastoral setting but also made a deep scar on the Indian traditional agrarian culture and the livelihood of people. Second, it is a saga of the Third World women peasants' resistance to survive against modernisation and corporate agriculture. Third, the novels show how agriculture is not only a major occupation of Indians but it is also a means of subsistence for survival along with their divine reverence to nature. Along with these aspects of ecofeminism, these novels add a new dimension to the ecofeminist discourse through their unique vision. Each novel shows two unique aspects: man is also the nurturer and subjugated by the development attitude. The fact that man is a saviour of nature is shown through certain characters: Nathan in *Nectar in a Sieve*, Abu in *The Upheaval*, Rama the grandson of Aithala in *Return to Earth* and Dinakaran in *Gift in Green*. The other concept is that man is also subjugated by the development attitude is evident in *Nectar in a Sieve* through Nathan and his sons, in *The Upheaval* through Pandhari and Nanu, and in *Gift in Green* through Dinakaran. The reversal of conceptual thinking demonstrates the ideal shift – ecofeminism to ecohumanism – which is a practical reality of India.

Note

1 The current section is the derivation of "Reconstructing ecofeminism: A study of Kamala Markandaya's *Nectar in a Sieve*" by Patil Sangita Sharnappa, used under CC BY.

Bibliography

Cuomo, Chris. *Feminism and Ecological Communities*. Routledge, 1998.

Daly, Mary. *Gyn/Ecology: The Metaethics of Radical Feminism*. Beacon Press, 1978.

Datar, Chhaya. *Ecofeminism Revisited*. Rawat Publications, 2011.

D'Eaubonne, Francoise. "Feminism-Ecology: Revolution or Mutation?" *Ethics and the Environment*, vol.4, no.2, 1999a, pp. 175–177.

———. *Le feminism ou la mort*. Pierre Horay, 1974.

———. "What Could an Ecofeminist Society Be?" *Ethics and the Environment*, vol.4, no.2, 1999b, pp. 179–184.

Foss, Karen, et al., editors. *Readings in Feminist Rhetorical*. Sage Publications, 1999.

Hemenway, Stephen. *The Novel of India*. Inter Culture Associates, 1975.

Joseph, Sarah. *Gift in Green*. Translated by Valson Thampu. Harper Perennial, 2011.

Karanth, Shivram. *Bettada Jeeva*. Sapna Book House, 2014.

———. *Chigurida Kanasu*. Sapna Book House, 2014.

———. *Chomana Dudi*. Sapna Book House, 2010.

———. *Marali Mannige*. Sapna Book House, 2012.

———. *Mukajjiya Kanasugalu*. Sapna Book House, 2012.

———. *Return to Earth*. Translated by Padma Ramachandra Sharma. Sahitya Akademi Centre for Translation, 2002.

———. *Sarasammana Samadhi*. Sbs Publishers Distributors, 2015.

King, Ynestra. "The Eco-Feminist Perspective." *Reclaiming the Earth: Women Speak Out for Life on Earth*, edited by L. Caldecott and S. Leland. Women's Press, 1983a.

———. "The Ecology of Feminism and the Feminism of Ecology." *Healing the Wounds: The Promise of Ecofeminism*, edited by Judith Plant. New Society Publishers, 1983b.

———. "Engendering a Peaceful Planet: Ecology, Economy, and Ecofeminism in Contemporary Context." *Women's Studies Quarterly: Rethinking Women's Peace Studies*, vol.23, no.3/4, 1995, pp. 15–21.

Markandaya, Kamala. *Nectar in a Sieve*. Penguin Books, 2009.

———. *Pleasure of City*. Penguin Books, 2012.

Merchant, Carolyn. *The Death of Nature*. HarperCollins Publishers, 1980.

———. *Earthcare: Women and the Environment*. Routledge, 1995.

Mies, Maria, and Vandana Shiva. *Ecofeminism*. Rawat Publications, 1993.

Mukherjee, Upamanyu Pablo. *Postcolonial Environments: Nature, Culture and the Contemporary Indian English Novels*. Palgrave Macmillan, 2010.

Naik, Pundalik. *The Upheaval*. Translated by Vidya Pai. Oxford University Press, 2002.

Ruether, Rosemary Radford. "Ecofeminism: First and Third World Women." *American Journal of Theology & Philosophy*, vol.18, no.1, 1997, pp. 33–45.

———. *New Women/New Earth: Sexist Ideologies and Human Liberation*. Seabury, 1975.

Sangita, Sharnappa Patil. "Reconstructing Ecofeminism: A Study of Kamala Markandaya's *Nectar in a Sieve*." *Cogent Social Sciences*, vol.2, no.1, 2016, pp. 1–17.

Selvamony, Nirmal. "Considering the Humanities Ecotheoretically." *Journal of Contemporary Thought*, vol.40, 2014, pp. 5–19.

Selvamony, Nirmal, et al. *Essays in Ecocriticism*. Sarup & Sons and OSLE-India, 2008.

Shiva, Vandana. *Biopiracy*. Natraj Publishers, 2012.

———. *Staying Alive Women, Ecology and Survival in India*. Women Unlimited, 2010.

———. *The Violence of the Green Revolution*. Zed Books Publication, 1993.

Starhawk. "Feminist, Earth-Based Spirituality and Ecofeminism." *Healing the Wounds: The Promise of Ecofeminism*, edited by J. Plant. New Society Publishers, 1989.

———. *Webs of Power: Notes from the Global Uprising*. New Society Publishers, 2002.

Takle, Niranjan. "Digging One's Ore Grave." *The Week*, vol.30, no.41, 7 October, 2012, pp. 33–37.

Tong, Rosemarie Putnam. *Feminist Thought: A More Comprehensive Introduction*. Westview Press, 2009.

Warren, J. Karen, editor. *Ecofeminism: Women, Culture, Nature*. Indiana University Press, 1997.

———. "Ecofeminist Philosophy and Deep Ecology." *Philosophical Dialogues*, edited by Nina Witoszek and Andrew Brennan. Rowman & Littlefield, 1999.

———, editor. *Ecofeminist Philosophy: A Western Perspective on What It Is and Why It Matters*. Rowman & Littlefield Press, 2000.

———. *Ecological Feminism*. Routledge, 1994.

3 Dam construction and ecological crisis

The Coffer Dams and *Dweepa*

3.1 Introduction

In the previous chapter, by examining four novels, we studied the impact of modern developmental attitudes on the agrarian and pastoral life. We tried to read select novels from the point of view of agriculture as an ecofeminist concern. The novels discussed in the previous chapter show the concern for environmentalism, and largely their concern is erosion of agriculture due to modern development markers. The present chapter examines Kamala Markandaya's *The Coffer Dams* (2008) and Na D'Souza's *Dweepa* (2013) as environmental narratives that draw our attention to the impact of dam construction, another aspect of modern development, on tribal people and their idyllic life. These novels document the agony of displaced local people due to the submersion of land. Before undertaking the analysis of these novels, let us understand the project of dam constructions in India as a developmental programme of the state and other agencies.

3.2 The project of dam construction and ecological crisis in India

In terms of the number of dams, India stands fourth at the level of global ranking. There are more than 4,000 dams in India. Most of the dams in India are constructed and maintained by the Water Resources Departments and the state governments. The Planning Commission has been discussing inclusive development, while the Ministry of Environment and Forestry and the Ministry of Rural Development have been drafting and redrafting policies regarding the impact of dam construction on environment and resettlement of people. However, the framed policies are not up to the mark or not pertaining to the needs of the displaced people (Scheumann and Hensengerth 2014:96–97). Ramchandra Guha, a noted Indian historian and an environmentalist, says that the major focus after independence was on technology-driven industrialisation, which was considered as Nehruvian era (Guha 2006). Pandit Jawaharlal Nehru, the first Prime Minister of India, observed that dams are "the temples of Modern India," and thereby referred to many of their benefits: to store rain water, irrigate farmland, generate electricity for industrial use, supply drinking water, and save land from floods and draughts (Cullet 2007; Scheumann & Hensengerth 2014).

But, this is simply one side of the coin; the other side has a horrendous impact on environment and human beings. In July 1983, the veteran social worker Muralidhar "Baba" Amte wrote to the Prime Minister, Indira Gandhi, urging her to intervene in stopping two dams in central India that would submerge 2,00,000 acres of dense forest. These dams would also displace 40,000 adivasis; although they would be paid monetary compensation, "nothing can compensate for the wrench they would suffer in leaving their traditional cultural environment." In terms strikingly reminiscent of Nehru's 1958 address, Amte argued that "it might not be necessary to incur the multiple costs and risks in building more dams of gigantic size." "A series of small dams," he continued, could "adequately meet the water and energy needs of the people, including electricity for industry, without degrading the environment." On July 18, 1984, Indira Gandhi replied to Amte's letter that she is "most unhappy that development projects displace tribal people from their habitat, . . . sometimes there is no alternative and we have to go ahead in the larger interest" (Guha "Prime" 2005).The basic and major problem is displacement or rehabilitation of tribal people who lose their lands, homes, livelihood, property, age-old culture and tradition. The indigenous or tribal people are not only deprived of their culture and kinship activities but also their whole socio-economic and ecology-based texture along with their intimate relationship with nature. They lose their inherited knowledge of plants and animals of that area, which is their source of livelihood. Despite their considerable degradation, the forests continue to be an important life support system for the people in the submergence zone, and still contain a diversity of flora and fauna. In addition, severe environmental impacts are anticipated downstream of the project and even outside the so-called impact zone (Agarwal 1998). It leads to further confrontation and multifarious agglomeration of ecological crisis by shifting valuable biological and cultural diversity.

Dams, according to a prominent Indian ecofeminist thinker and activist Vandana Shiva, are detrimental for a number of reasons, including the fact that "diversion of water from its natural course and natural irrigation zones to engineered 'command' areas leads to problems of water-logging and salinity. Diversion of water from its natural course prevents the river from recharging groundwater sources downstream" (Shiva 2010:185–186). The fundamental objective of the construction of dams is to fulfil the demands of a handful of urban people, which is considered to be 'development' in 'national interest.' But, the local people have to sacrifice their basic needs and livelihood. We can discuss this argument through Mies' and Shiva's criticism:

> It was only during the 1980 when the different 'local' interests met each other nationwide, they realized that what was being projected as the 'national interest' were the electoral and economic interests of a handful

> of politicians financed by a handful of contractors and industrialists who benefit from the construction of all dams such as Tehri and the Narmada Valley projects. Against the narrow and selfish interest that had been elevated to the status of the 'national' interest, the collective struggle of the communities engaged in the resistance against large dams started to emerge as the real though subjugated common interest.
>
> (Mies & Shiva 1993:10)

Against this short profile of the project of dam construction in India, we will proceed to examine how the two novels critique the project and show their ecofeminist concern.

3.3 Kamala Markandaya's *The Coffer Dams:* modern juggernaut

In the previous chapter, we examined *Nectar in a Sieve*. After more than a decade after her first novel, *Nectar in a Sieve* (2009), Kamala Markandaya once again dealt with an ecological crisis in *The Coffer Dams* (2008), especially focusing on the impact of dam construction on the tribal people and nature. *The Coffer Dams* (2008), Kamala Markandaya's sixth novel, is an instance of a classic defence of ecofeminism. The novel is a saga of exploitation of tribal people and nature by a patriarchal modern developmental attitude in the form of dam construction.

Kamala Markandaya's *The Coffer Dams* (2008) gives a broader perspective on the construction of dam and its adverse effects on tribal people, animals and nature. *The Coffer Dams*, as the title itself indicates, revolves around the construction of a dam that was a living problem in post-independent India, that is, the Nehruvian era. This period was considered a developmental era, which in turn has created ecological stress and a variety of social problems in the Indian agrarian and indigenous culture. The novel explores the dynamics of cultural imperialism towards nature and human beings. In a way, it is a deep and abiding tale of perpetuated dehumanising anthropocentric attitude, which underpins different insulators of progress such as industrial civilisation and development, colonisation as well as capitalism, with its focus on the construction of a dam that is an offshoot of the development project.

The Coffer Dams has multifarious trajectory of themes with more value being attached to the earth and organic elements, the primal kinship between man and nature on the one hand, and, on the other hand, they also show sincere empathy with the dark side of human nature as interpreted in terms of manipulative and exploitative manoeuvres of Western thinking. The novel enunciates the impact of blatant disregard of a group of British and Indian technocrats, who try to build a dam in a hilly tribal region in Southern India. The British engineering firm partnered largely by Clinton and Mackendrick along with Indian engineers like Krishnan, Gopal and local technicians like Bashiam and labourers, try to tame nature by civilising it. On the other hand, the old tribal man always worries about his men and forest. He wants to save his people and forest from the modern juggernaut, a glamorous world that takes its toll on the forest and their livelihood.

Clinton, the protagonist, always delineates his supremacy over Indians as well as on nature. Mackendrick, another character, is eager to win over natural phenomena like the monsoon cycle and cyclones for the construction of the dam. When his efforts go wasted despite using different techniques he feels disappointed, whereas Clinton collects a lot of data about the North-West monsoon. However, monsoon becomes uncertain because of men's destructive nature. Greed pervades everyone: Bashiam, a tribal man from the hills, skilled in the traditional craft of woodcutting, is attracted to the Whites. The working of strange powerful turbines and incentives given to his work fascinate him further. This glamorous world not only uproots Bashiam but also many of his men from his family, ancestral professions and their village.

3.3.1 The effect of dam construction on an idyllic hill

The Coffer Dams (2008) is a deep and abiding tale of a perpetuated dehumanising androcentric attitude. Major ecofeminists' propound that the base of an androcentric attitude is industrialisation, development, colonisation and capitalism. This novel deals with the construction of a dam that is an offshoot of development. The narrative provides an opportunity to ponder over the effects of dam construction on tribal people and nature.

The dam construction is a metaphor for the developmental patriarchal attitude. To explore this concept, we have to analyse some major characters. The British engineers, Clinton and Mackendrick, are the embodiment of colonial power and Westernisation. Further, the Indian engineers and technicians like Krishnan, Gopal and Bashiam join their hands with Clinton and Mackendrick in the project of dam construction. These British and Indian Engineers and technicians exhibit the feeling of strong determination to construct a dam by cutting down trees, encroaching upon the habitat of animals and uprooting tribal people from their culture. Clinton, the protagonist, always exhibits his supremacy over Indians as well as on nature. Clinton and Mackendrick are eager to win nature's natural phenomena. Clinton collects a lot of data of the North-West monsoon from the records of 300 rain gauge stations. The basic intention of this meticulous study of monsoon is to forecast that year's monsoon by which he can continue the dam construction without any delay. However, the monsoon becomes uncertain because of men's destructive attitude.

The novel opens with an in-depth description of the town, a concrete jungle in a natural jungle, which is built by Clinton. The town is a man's town and built by a man. It also focuses on the consumer culture or modern culture, which is plagued by a materialistic world such as "a Coffee Club," "a Soft-drink stall" and "a tin Shack." The road on which these stalls are installed is named as Clinton's Line. These stalls and clubs are an emblem of commerce brought into the hilly forest area in the midst of a natural setting. The ecofeminist tension of the capitalist patriarchy against wilderness is noted by Mellor, "The dilemma of human embodiment exists as a fundamental feature of the human condition but it becomes most destructive in the divided societies of capitalist patriarchy where domination and transcendence of the natural world is central" (Mellor 2000:111). The tribal people have lost their eco-friendly houses and shifted to temporarily constructed houses that "there were those long low structures, row after row, like old-time barracks, where the workers lived" (Markandaya 2008:1). The living conditions of tribal people portray the adverse effects of the construction of dam on them. The novel deals with the tension between the tribal lives and the construction of dam in serene natural land.

Clinton always feels very proud because he is always praised as a man of wealth, property and success. The proud feeling of a male, a builder of The Great Dam, is the mark of destruction that is rampant in these days. He always considers himself as a highly polished, perfect and technically skill-oriented builder: "A builder. The word ran through his mind with a dear keen pleasure" (2). For Clinton and Mackendrick, it is a very tough time to get sanctioned the tender of dam construction in the prevailing cutthroat competition. Second, in spite of many difficulties posed by climate and terrain to construct the dam, Clinton prepares a preliminary blueprint of the construction of dam with the assistance of Henderson, Mackendrick, Rawlings, Lefevre and Galbraith. As per the blueprint, it takes a year to construct a diversion channel to alter the course of the river where they plan to build the coffer dams and two years to stem the main dam if the monsoon co-operates with their plan. Clinton finalises the plan along with the British senior officers by considering the South-West monsoon. Clinton is always obsessed and preoccupied with the dam construction and he always sings the plan of dam construction as a psalm.

There are various phases of dam construction that takes a heavy toll on nature. As the first phase of dam construction, Clinton and Mackendrick start the project by constructing a good deal of bungalows with gardens, gravelled paths and whitewashed boulders for senior officers, "town houses in woodland setting" (7) and also quarters for workers. In addition, Mackendrick constructs a road because Clinton expects the arrival of an exported Ferrari 500. Second, to construct the great dam, they erect a small industrial town, install a workshop, loading and unloading bays, car maintenance sheds, workers' quarters, engineers' bungalows, amenity buildings, a water tower, ice and filtration plant, pumping and power stations. Subsequently, the construction of dam goes on quite reasonably by continuous daily blasting and removal of debris by cranes and lorries.

The emergence of men destructive nature – considered as the agent of development – has torn apart the jungle idyll. It may be observed here,

> The Great Dam, it had come to be called: not by him, he was too absorbed by the work in hand for adjectival excess, but by people of the Maidan and

> the Malnad, the plains and hill-country people, who had watched with awe the precipitate birth of a town in the jungle. It was virtually a small industrial town, gouged and blasted out of hillside.
>
> (2)

It grabs the livelihood of tribal people and habitat of animals for the benefit of urban people. The coffer dams rise to block the course of the river and divert the water to channel it downstream by which the river will resume its natural flow without disturbing the construction of the main dam. About his ambitious project, Clinton opines for "this river which two thousand men and ten thousand tons of equipment had so far assembled to tame" (29). Clinton and Mackendrick start their preliminary phase by working on the plateau to split it open with dynamite to create the channel into which the river would flow 20,000 cubic feet per second. The patriarchal attitude is very clearly depicted through the words of Mackendrick: "When river's in spate. A freak spate isn't calculable, and freaks aren't entirely off the cards. The river's the real bastard" (30).

Next, the men and the machines have been working day and night to get the construction work done in time. Maintenance sheds are installed on an acre of land by clearing the dense forest. It has expensive equipment and many men are appointed to repair and operate these machines. The next progress of the construction of dam is the explosion of dynamite. Everyday 20 explosions by using 25 tons of dynamite take place. Misty and dusty clouds cover the entire vicinity. Later, the excavation work starts in the south of the main area, where the riverbed lies, by drilling, shovelling and chiselling the rocks. In order to make the way to water, trucks move continuously and heavy machines work with the labourers to clean the debris and the dump. Five tons of debris every two minutes is removed to create a channel to change the course of water by which they can build the coffer dams. The daily course of life in the jungle is pounding, blasting and drilling, which might be the habitat of many animals and species, but everything now looks misty in the dusty clouds of construction. One can, for a while, pause and meditate to what extent the process of dam construction pollutes the air, water and land. Even it has grabbed the habitat of many animals and birds along with the tribal people's livelihood and their kinship with nature. This is an uncomplimentary critique of the modern attitude of man.

Further, Clinton and Mackendrick continue their efforts to collect the equipment and materials such as bulldozers, excavators, pumps, tyres, dumping trucks, barges and high-load capacity cranes to complete the construction in time. The premature blast in the valley engulfs 28 people and their families. Clinton is least bothered about this disastrous incident. He doesn't have in him the milk of human kindness. He is engrossed in his objective of the dam construction: "This with the bleakness of the crags that surrounded them upon his face, battling with them to prevent them crushing, closing in before what lay in his mind could grow, accumulating power and inevitability" (171). His struggle and effort to build the dam makes him forget the value of human life for a while and poses the question – Is he a man or a machine?

Let us pay attention to other prominent characters such as the Indian engineers Krishnan and Gopal and the technician Bashiam. The Indian engineer Krishnan, the leader of the Indian delegates, puts forth a picture of the Indian monsoon that is unpredictable as well as labour problems which are endemic. Therefore, he suggests that they cannot prepare a tight schedule for the completion of dam construction. Later, he adds that it is not only the delay in the completion of dam but also the reputation and prestige of the government at stake. He wants to modify the blueprint of the construction of the project. Markandaya exhibits the universal patriarchal attitude by portraying the attitude and mentality of Krishnan, "We are on the same side, aren't we? We're working for the same thing and we ought to be working together and he knows it as well as I do" (16). Krishnan always thinks he is superior like Clinton with his 'imperial insolence,' and he is aloof from the other workers. Krishnan, in spite of his being an Indian and though he is very much acquainted with the Indian angst and anxieties, is indifferent to the tribal people and their settlement. Being Westerners, Clinton and Mackendrick may feel nauseated by the tribal people's lifestyle and their ignorance. They feel that these tribal people have "No shame, no dignity." However, we can see the same attitude in Krishnan, as he too feels the tribal people are like cattle: "Look at them! Lined up passive cows at backstreet Christian butchery!" (68). In spite of being an Indian, he has no empathy for the tribal people and their problems.

Markandaya has sketched many characters to represent the patriarchal developmental attitude. However, we will focus mainly on two characters who depict this ubiquitous feature of the West and India through Clinton and Bashiam. These characters represent the Western and Indian perspectives of the developmental attitude, which is the key feature of ecofeminism. The entire novel is occupied with two major characters, Clinton and Bashiam, who support dam construction. Bashiam, a tribal man, represents a few tribal people's attitude towards dam construction. Greed prevails everywhere; Bashiam, skilled with the traditional craft of woodcutting, having inherited the knowledge of the forest, the river, the seasons and the hill, gets attracted to the project of dam construction. The working of the strange powerful turbines and the incentives fascinates him. This glamorous world uproots him from his family, ancestral occupation and his village. He starts learning how to repair machines. In spite of having his roots in the tribal culture, he is attracted to the construction of dam and modern juggernaut. On the basis of his little technical knowledge without having any formal education, he joins the project of dam construction. Further, he learns skills like building, repairing, dismantling and welding the machines in association with the British foreman and other workers.

Clinton considers him as a *Jungly wallah*: a man of the jungle, a savage or a primitive man as if he just came down off a tree. Markandaya very minutely elaborates that there is a paradigmatic shift in the perspectives of Bashiam.

Bashiam forgets his culture, values and people; he hardly likes to visit his huts; he starts living in the room provided by Clinton for the workers of the dam construction. Bashiam thinks that he no longer belongs to the tribal huts. Along with Bashiam many more tribal people have given up their native occupation and have been attracted to the wealthy buildings and contracting firms. Markandaya writes, "Bashiam's roots were attenuated: his homecomings were uneasy surface affairs, but not given to dwelling on lacks and losses he made do. And in the camps while the work lasted it was good. What it was he could not easily say. A feeling. A richness" (44).

Bashiam continuously operates, with complete devotion, the crane and the machine in order to lift boulders. In spite of exhaustion and tiredness of working on the crane from early morning to late evening, he, like Clinton, becomes obsessed with completing the dam construction. His harsh way of answering to the people shows his anxieties for the completion of the project: "The monsoon has not started. . . . Only the preliminary rains. There is still time in which to finish" (139). He knows very clearly that his response to the people's remark is not factual and even he knows about the uncertainty of monsoon rain. Yet, it is a simple attempt to presuppose that they are able to complete the project of the dam construction in time. Bashiam, with the help of the crane, traps the corpses of the labour, who are victims of a premature blast. During the course of lifting the boulders and the corpses, he meets with a major accident because of machinery defects, and although he escapes death, he is bed-ridden for several days.

3.3.2 Women and nature: an intimate bonding

The second trope of ecofeminism is women as nurturer of nature. Francoise d'Eaubonne, the founder of the concept ecofeminism, very aptly remarks, "Worldwide, women are more involved in the ecological problem than men. The grounds: they give life, and have, therefore, more of the concern of future generations" (d'Eaubonne 1999a:176). Now, we will try to study the feminine qualities of caring and nurturing through Helen. Helen, the central character of the novel and the wife of Clinton, always has an intimate association and bonding with the tribal people and nature. Clinton brings Helen to that remote land. His impression is that it is very difficult for Helen to adjust to that primitive frontier town, as they are roughly hewing into existence. This typical attitude demonstrates superiority, but Helen never feels that she is in an alien land. She feels more comfortable in the association of the tribal people, animals and nature's beauty. She always wanders all alone. Markandaya would like to picture her concern, affinity and kinship with the tribal people and forest, "Helen got on well with the tribesmen. He had seen groups of them gathered around her in their compound or accompanying her if she returned after sunset from her wanderings. But then, Helen got on well with most people, even housewives whose sorties down the hill on club or shopping expeditions she never joined" (Markandaya 2008:21).

Markandaya very delicately interweaves the two core tropes of ecofeminism – the two juxtaposed attitudes – that is the patriarchal developmental attitude and women's concern for nature. Helen brings and shows small pieces of pottery and very inquisitively asks Clinton for the reason why those pieces are lying in the compound. Clinton explains, "The locals, I mean . . . some of 'em were camped here before we moved in, I'd quite forgotten that little episode" (23). It elucidates men's indifference to other human beings and nature. In addition, he says that the tribal people were asked to move from that place because that site would be needed for the construction of the dam. It is the prime site and an ideal place for them. Further, he thinks those are only huts, which can be easily evacuated. Clinton fails to understand the real meaning of the huts for the tribal people. He says that these tribal people were moved from that place like animals. The local people are uprooted from their native culture and place as the result of dam construction. The episode doesn't bother Clinton, but the same episode leaves a deep scar on Helen's heart. She contemplates how the local women might have cooked food in the earthen potteries – which are now in pieces – and served their families:

> It had been part of some woman's life once, not very long ago: she had filled it with water and scoured it, cooked in it and fed her family – the earth ware was pebble – smooth from use. Then they had all gone away and the vessels had been broken and left behind. Not one or two: enough for several families, the cooking pots of a whole cooking pots of a whole community.
>
> (24)

This episode of the broken pieces of pottery symbolises a wide ramification. It evokes several questions in her mind. It stands testimony to modern civilisation and the patriarchal developmental attitude, which sweeps away the tribal people from their domain into the darkness by replacing it with bungalows. Clinton never considers an Indian as a human being; in contrast, Helen develops a sense of intimacy with and sympathy to the tribal people. Though they belong to the same land, there is a yawning gap in their thinking. In this context, Laxmi Kumari Sharma's remark is very apt, "Helen considered them human beings and overcame the racial burden of getting beyond the skin. She thinks that Clinton lacks certain essentials of human beings. Hence, she drifts away from him" (Sharma 2001:24).

After the premature blast, many people meet with death and when the bodies of two Indians cannot be traced, the tribal labourers go on strike. Clinton and the other British senior officers are not able to understand the reason behind the strike. Especially, Clinton is very indifferent to the tragic incident and the death of many people. He is only preoccupied with the project of dam construction. But, Helen can understand their pulse and has a kindred soul for them, "But these people aren't different clay, they're like me, like people like me. What is for me is for them, there's no other kind of yardstick that's worth anything" (Markandaya 2008:45–46).

She has a fellow feeling and benevolence not only for the tribal men, women and children but also for animals. She rubs flea powder on dogs and helps the dogs get rid of parasites. Helen always supports to tribesmen and Bashiam. She is against of the construction of dam. She posed a question to Bashiam to protest against the inhuman treatment to the tribal people as, "You were – you are – a member of that tribe. It was their land. They didn't want to leave it, they were persuaded. Why did they allow themselves to be? Why did you? Without even protesting?" (46).

Markandaya portrays the core ecofeminist concern in her novels. Further, we can examine a few factors that focus on ecological concern.

3.3.3 Deep reverence for nature

For the Indians, natural resources are not only means of survival, but also they have a deep reverence for these resources beyond their material benefit. In this novel, we can trace that men and women both have an equal reverence towards nature. They think that nature is their deity, "But what weigh if any, he thought with contempt, could one attach to the words of a people who worshipped birds and beasts and probably snakes, decking the forest with scruffy hutches which they knocked up out of driftwood and crammed with leaves and flowers for their deities?" (76). The tribal old headman very silently contemplates the activities of the construction of dam. Being ignorant of their base, these tribal people join their hands with the white people in the dam construction. The tribal headman regrets the choice of his people going against the natural phenomenon and joining the modern juggernaut. He ponders, "Well, malice was a better safety vent than many for the heartburning his men must have caused him, defecting from their duty to him to themselves, which he must have felt for all his serenity, the clam whose priceless grains had been meticulously isolated over a lifetime" (152).

Krishnan's affinity and belongingness and love for his traditional occupation and his roots can be analysed by his remorse for the degradation of the pasture land, which is the victim of "the modern juggernaut." This phrase is repetitively used in the novel in order to emphasise the impact of introduction of dam construction in the serene land. His past sweet reminiscences include: "I suppose all this must remind you of your precious paddy fields. . . . All these wet acres just like paddy fields, for which you have thirsted ever since you came to these hills" (217). Krishnan engrosses himself in his activities of the construction of officers' bungalows in the highland. Due to these modern activities, many small streams go drained and the entire idyllic hill turns into dry views. He believes all men drown in the well of loneliness. In addition, we can see his conviction and confession, as a wrong doer, "To me it resembles homicide" (231).

3.3.4 In-betweenness: modern project and native roots

The major element of the ecofeminist discourse is the developmental attitude that is considered to be a patriarchal attitude because it is introduced and supported by men. However, the Indian novelists, along with the basic ecofeminist elements, have added a new dimension to the ecofeminist discourse by investigating a few men characters' in-betweenness. They are neither able to accept modern projects from the bottom of their hearts by cutting down their roots from their nativity nor reject them outright. This concept can be explored in Bashiam. He, a tribesman, is very much contented with the natural things like hills and woods. However, the introduction of dam construction leaves him no other alternative for survival but working on the dam construction. Bashiam oscillates between the project of dam construction and his ancestral occupation:

> Bashiam seldom paused to analyse his situation. He knew he no longer belonged in the tribal huts of his birth, but his brief returns to them he did not find too irksome. Equally there was no sense of belonging to what they called the jiffy towns, the tin and canvas camps the contractors rigged up at breakneck pace for their labour, but he did not look for it.
>
> (44)

Bashiam joins the construction work of dam. As per the eminent ecofeminists, it is an epitome of patriarchy. But, here we have to ruminate over a set of questions: is it his natural instinct? Or, is it out of his curiosity? Or, is there any inevitability to join the project? This aspect is very vividly depicted here. "Bashiam was an outsider – de-tribalized, he had heard them say of him. Sometimes, without undue agonizing, he acknowledged the truth of it. He also knows in his bones that however de-tribalized he might be birth and upbringing within the tribe gave him race knowledge and instincts that could never be acquired by the real outsiders, those who had never been inside" (81). Bashiam is neither able to be a tribal man nor a modern fellow. Most of the time he shows his modern attitude but everything is temporary and superficial. His craving for the hut shows, on the other hand, his deep bonding with his nativity is "no longer belonging, a man put his shoes on his feet and worked machines, whose feelings and desires they could not fathom" (135).

The entire novel projects Bashiam's continuous struggle to come out of his in-between-ness. This analysis adds one more dimension to the ecofeminist discourse that man is not only a destructor but also a saviour of nature. Next, we evince the mental encounter of Krishnan and Gopal – the Indian engineers and technicians with the British engineers (Clinton and Mackendrick). The apparent purpose of these two is to complete the dam construction in time, which is a symbol of pride. Nonetheless, when we probe into Krishnan's and Gopal's inner feelings and intimacy with the river, we get one more layered meaning: "We are river people, or, I would say, an inland people who live by our rivers" (158). They incline to the river, the forest and the tribal people. The tribal men are not ready to work on the project of dam construction because of the death of 28 men. Moreover, a few dead bodies are not traced. They demand that until they get the dead bodies of their kinsmen, they are not ready to resume work.

Though Krishnan is very much preoccupied with the completion of dam construction, his predisposition or concern is more inclined towards the tribal people and their problems. He says that he is very much with them and ready to support them in their thick and thin because they are the offspring of the same soil. A minute examination of the characters like Krishnan and Gopal show their condition of in-betweenness. They are neither able to be a part of their ancient culture nor completely accept modernisation.

In addition, the novel throws light on yet another quality of these men. The Indian engineers and technicians like Krishnan, Gopal and Shanmugham and the tribal man Bashiam – every one of them – are committed to their work of the dam construction along with the British engineers and technicians by their body, mind and spirit for the common end. However, their ultimate appeal is to support their native people and culture. Therefore, the novel highlights that though all these Indian engineers, technicians and workers have triumphed over their age-old profession by supporting the dam construction, their inner call is always with nature and the tribal people. In reality, these tribal people are simple, nature lovers, real developers and eco-friendly.

3.3.5 Ecological sensibility

Markandaya shows the Indian biodiversity through various kinds of birds (viz. sunbirds, bulbuls, finches, hill mynahs, kingfishers) and other animals. Second, the tribal people are eco-friendly by their lifestyle, for example, living in the nature friendly huts, wearing palm leaves as cloaks and using jute mantle and Palmyra umbrellas. Thus, Markandaya has given very minute details of Indian forests, hills and life of tribal people along with the developmental attitude. The entire novel is infused with ecological sensibility and the reciprocal relationship between man and nature. The entire gamut of the novel weaves a fabric around the idyllic hill.

3.3.6 Shortage of resources: impact on livelihood

The headman's contention and agitation about the dam construction in the prime location, which is the tribal people's settlement, makes him rethink about tribal people's future. "He saw the dust from the dams like as on his tribesmen's faces, and the growing neglect of the village as more and more of his men were sucked in, to whirl like cogs around the restless core" (150). He knows about rain; the rise of the river; each day how the water of the river accelerates. However, he feels helpless about the panicky condition of lowlanders who have come up the hill to collect their daily wages. They lose their livelihood and for them the construction of dam is the only alternative for their survival. He is very much worried about his men who are involved in dam construction. Always, he thinks that their ancient profession is based on nature and these kinsmen forget their dependency on nature and the role of nature in their lives.

Further, Markandaya very minutely portrays the mental anguish of the headman who gapes at the deterioration of nature and their men. "He could, if he wished, have exerted his authority, compelled unwilling men to return to a skimped care of the village – the elemental care which was as much as it had ever received or would receive at the hands of its harried people. He did not" (151).

3.3.7 The uniqueness of The Coffer Dams *as an ecohumanist critique*

The novel, besides presenting the essence of ecofeminism, discovers that Indian environmental problems are not gender oriented but are gender neutral. Therefore, men too are saviours and subjugated by the development attitude along with women. Let us scrutinise these elemental aspects. The novel ruminates on the struggle between the civilised and the uncivilised world. On the one hand, civilisation is introduced in the form of dam construction by a group of British engineers, senior officers and workers along with Indian engineers. On the other hand, the novelist projects uncivilised people – peasants and tribes – who have deep concern about nature. This concept is scrutinised through various characters, rhetoric modes of presentation and literary devices.

We can say that the destruction of nature is not an androcentric attitude but an anthropocentric attitude, that is, human/ nature. Moreover, man is not only destroying nature but also he is a saviour of nature. Let us discuss a couple of characters in support of this argument: the tribal old head man, Bashiam, Krishnan and Gopal.

The tribal old man is very much disgusted with two things. First, the project of dam construction on a serene idyllic hill has taken a toll on the habitat of many animals and tribal people. Second, he is not able to resist the involvement of the tribal people in the construction of dam. He shows his reluctance: "They are short of food too, whose fault is it, the jungle of game, if they relied on that and not on the money which comes and goes – but what is the use of an old man talking. Keep away, I told them . . . I am their headman. I have to say these things; someone has to say them; but no. Now they are punished and are hurt, like small children. Like fools whose faults" (72). It is very apparent that he neither goes against the order of nature nor is in support of civilisation and modernisation. Therefore, his fervent appeal is to be very much amidst nature and a part of nature. As modernisation, development, scientific revolution, colonisation, rationalistic culture and global free market are considered to be the outcome of the patriarchal attitudes which are the root causes of exploitation of woman and nature (Shiva 2012; Mies & Shiva 1993; Merchant 1980; Eaton & Lorentzen 2003). We can't wholly withstand the above-mentioned Western ecofeminists' ideology in relation to a few men characters presented by the Indian novelists. The worries of the tribal old man reorient the consequences of many tribal people's decision to support and work for the dam construction. It is a temporarily glamorous world and may give monetary benefits but they fail to understand the long-lasting effects on their occupation, culture and nativity. The old man describes, "A peace full of moaning and pining for trash. But before that they will learn what is real and mourn what is lost. A score or more before they bend the river . . . the Great Dam will take them, the man eater will have its flesh" (72).

We espouse the same aspect through Bashiam, a hill tribesman, who joins the dam construction as a technician. His new learning viz. building, repairing, dismantling and welding machines counterpart to his traditional craftsmanship of wood cutting, inherited knowledge of the forest, the river and the hills uproots him from his native bonding, village and family. However, his inner soul always inclines to his assimilation into nature. "To his tribe he was a man who walked alone, sprung from them but no longer belonging, a man who put shoes on his feet and worked machines, whose feelings and desires they could not fathom" (135). Further, Bashiam's suffocation at mental and physical level is portrayed. "For a while he lay on his bunk, staring up at the ceiling which oppressed him, white and blank a winding sheet. He switched off the light and it was worse, the whiteness precipitated itself closer, he felt if he breathed it would come down on him. At length he pulled himself up from clutching inertia, dressed, and went out" (136).

The same feeling that man is a part of nature is shown in another minor character Gopal, who assists Krishnan in the dam construction. He is against meat eating and animal slaughter. He never consumes flesh of animals. He thinks eating meat is a necrophiliac activity and unfitting for human beings. These are a few characters that illustrate that men are saviours and protectors of nature.

In this novel, due to exploitation of nature – compared to women – the men suffer more. "And the long lines of cheap labour could be seen, working ineptly, in a way that consolidated every atom of contempt in which Clinton already held them, alongside highly efficient costly machines, carrying away in shallow trays on their heads all manner of detritus, gravel, clay, the grey sludge from the river banks that oozed through the wicker and wattle on to their naked backs" (150–151). Further, the premature blast during the dam construction has taken a toll on 28 tribal men.

The headman is totally broken when he comes to know that the river engulfed 28 men. The loss of his kinsmen leaves him in irreparable trauma. His eyes are always wet because he loses his inner spirit. He is not able to disclose his agony of the exploitation of his men – the loss of their ancestral profession – and nature. Moreover, the old man is not able to withstand the blow of the loss of his men. Therefore, he is bed ridden. The novel gives an account of the anxieties of the tribal head man for the rootlessness of the tribal people and the destruction of nature. In this context A V Krishna Rao's views are very suggestive:

> Thus while the chief of the tribe symbolizes the suffering and tension as well as the undying hope and faith of the uprooted *autochthones* of the hills, Clinton represents the ruthless efficiency of the business-minded British engineers. The other characters such as Mackendrick and Krishnan sustain the dramatic tension in the novel. It is however, Helen and Bashim that symbolize the forces of moderation and human attitudes. They also impart a sense of urgency and intensity to the dramatic action and tragic development in the novel.
>
> (Rao 1972:40)

Thus, the novel apprehends the Indian environmental crisis, in the most compelling way, by providing glimpses of how men are first-hand victims of environmental disaster. Markandaya's ecofeminist concern may be found echoing in Murphy's remark, "Therefore, it is helping to envision a future that provides for the possibility of a more natured culture in which biotic differences and gender differences are celebrated in their diversity and heterarchy rather than used as justifications for domination, exploitation, and extinction" (Murphy 1997:57).

3.4 Na D'Souza's *Dweepa*: an island of destruction

Na D'Souza, a renowned Kannada novelist, has written 45 novels and many short stories. His two novels *Dweepa* (1978) and *Kaadina Benkei* (1988) have been adopted for feature films. Three of his novels, *Dweepa* (*Island* 1978), *Mulugade* (*Submersion*, 1984) and *Oddu* (*Dam/Barricade*, 1990), depict the tribulations of families because of dam construction. Displacement is the root cause of their woes and worries. Each novel delineates different aspects of displacement. *Dweepa* (*Island* 1978) recounts the impact of rising backwater on a family of Hosamane village, *Oddu* (*Dam/Barricade*, 1990) narrates the indifference of the government officers to provide compensation to families whose lands are above the full reservoir level, whereas *Mulugade* (*Submersion* 1984) shows their encounter with the new lives after leaving behind their age-old culture and land. No wonder, Na D'Souza is known as 'submersion writer' in the Kannada literary circles for his pre-occupation with the themes of dam construction and submersion.

Dweepa (2013) is divided into seven vignettes named after the stars that influence different phases of monsoon: specifically *Krithika*, *Rohini*, *Mrigashira*, *Aridhara*, *Punarvasu*, *Pushya* and *Aslesha*. Through the seven phases of monsoon, the novel traces the changing moods of nature. The project of dam construction on the Sheravati river displaces Hosamane village, which is inclusive of altogether five houses, among them three are landowners and two are bonded labourers. Three families are going to get compensation, whereas two families are not going to get any compensation because they are bonded labourers. Out of the three families, two are wealthy and bribe the officers, and so one of them that is Parameshwaryya, who gets the land near Sagara and the other Herambha who gets the land in Ananthapura. However, the family of Ganapayya doesn't get any land or money as compensation because he doesn't have any money to bribe the government officers. Therefore, he is left behind to face the consequences of submersion of his land. The novel narrates the plight of Ganapayya to survive amidst the problem of submersion.

Dweepa (2013) is an outcome of D'Souza's personal, keen and deep interest in the study of the problems of displaced people for which he spent about 25 years on a hydroelectric project on the Sharavathi River. This novel is a purge of his personal ethos, which shows the plight of community who are dislocated and uprooted from their native place. As the author says, "The tragedy in the lives of these innocent victims of modernization will now merge with the groans of the oppressed the world over, wherever this story is read. It is a fitting way to perpetuate the memory of those who lose all they hold dear wherever the country develops at their cost. It is also a proper way of mourning what we have lost because of what we have gained" (D'Souza 2013:xiii).

3.4.1 The displacement and the plight of native community

In comparison with the other select Indian novels for analysis, D'Souza's artistic narrative is very different. All other novelists portray ecofeminist elements like the modern developmental attitude through men characters. However, D'Souza showcases this basic trope of ecofeminism through crafty government employees who are greedy for money and through the very act of dam construction. Employees exploit innocent villagers by seeking bribes, harassing and cheating them in various ways. The ecofeminist stance is narrated in the matrix of dam construction, which forms the backdrop of the novel.

D'Souza constructs the narratives of community with the graphic presentation and aesthetic exuberance of the natural world. The fear of submersion of the land has been hovering over the villagers for more than a year. This can be seen through the threatening words of the elderly peon: "Sharavathi might swallow the Hosamane Parvatha this monsoon" (2). Another prophecy by the surveyors is that nothing would happen to them and they can stay there till they are compensated. The novel highlights the forthright ecofeminist stance by representing changes in the moods of the river, rain, land and the family of Ganapayya.

D'Souza projects the anguish and plight of submersion of land due to the construction of the dam through the male character Ganapayya. The very opening lines of the novel introduce Ganapayya – the villager of Hosamane – through two important concepts: the first, his dependency on and affinity with agriculture; the second, his age-old cultural practices. The fundamental reason to introduce these two concepts is to denigrate, due to the dam construction, how these two basic things are going to be disrupted in the near future. D'Souza not only desires to expound the exploitation of women and nature but also wants to depict the agony of local people. The villagers are not only deprived of their means of subsistence and survival but also uprooted from their occupation and culture. It is very suitable to throw light on the perspectives of

Mies and Shiva: "We also criticize the dualistic division between superstructure or culture and the economy or base. In our view, the preservation of the earth's diversity of life forms and of human societies' cultures is a precondition for the maintenance of life on this planet" (Mies & Shiva 1993:11).

Usually, the families of Hosamane grow areca nut on the farms and rice in the paddy fields. By producing a good harvest, they lead a happy life. But, because of the project of dam construction the two of the families along with two bonded labourers' families move to the other places. Only Ganapayya's family is left behind to face the ordeals of the project. He is not compensated with any land. Ganapayya puts forth the patriarchal developmental attitude exhibited by the government: "The government has set out to ruin thousands of homes. Is it a big deal for it to drown my village, my home? But what about the compensation they say they'll give us? When will that come?" (D'Souza 2013:3).

The most important thing they lose is their culture, which is more than 50 years old:

> The coconut and jackfruit trees in front of Herambha Hegade's house spoke of the antiquity of Hosamanehalli. His house was the oldest in the village. When Herambha's grandfather left a place near Mavingundi and bought a piece of land, built a house, and settled down here, his place came to be called Hosamane, the new house. Herambha's house became old but the village retained the name – Hosamanehalli – the village of the new house.
>
> (6)

Once, Parameshwarayya had an areca farm and paddy fields. After his displacement from the land, the field is full of weeds and shrubs. Further, the areca palms are chopped off to decorate the place where the minister is going to visit.

On the other side, Ganapayya's fields are full of tender finger-green seedlings. These are the two contrasting pictures that provide snapshots of the impact of dam construction on the land meant for cultivation. Ganapayya needs labour to work on his and Harmbha's fields to plant rice seedlings and cover the areca nut fronds. The labourers are not willing to stay on the land because they are afraid of what might happen to them during rains. Ganapayya used to go to Talaguppa – a city that is six miles away from Hosamanehalli – to buy their daily needs and groceries, and even to hire labourers to work on the farm by crossing the river. However, after the project of construction of dam, he is not able to cross the river like before. It is a tough time for Ganapayya to cross such a long distance by paying bus fare, and even it is too expensive to hire labourers. Further, no one is ready to work on their field, so the wages of labourers are hiked. It is inevitable for Ganapayya to pay soaring wages to complete the work – to tie palm sheaths around areca nut fronds, weed the fields and transplant the seedlings – before the rain starts. Ganapayya's family faces these consequences.

Next, D'Souza depicts the Indian eco-friendly ways of farming; that is, the organic and natural farming – mixed cropping – to retain soil fertility. Their lifestyle is also eco-friendly; they thatch their roofs with palm fronds and use banana leaves to have meals.

3.4.2 Woman's affinity and concern for nature

We can trace certain ecofeminist essence through Nagaveni, wife of Ganapayya. She feels isolated and alienated from the other people because no one is going to work on their land because of the fear of submersion. Earlier, Herambha's wife or Parameshwarayya's wife would come and discuss many things and the Hasalaru women, while passing by, used to talk of the happenings in their lives. Some women from Deervu and Aralagodu used to come and work on their farm, "Nagaveni would have felt there were some people around. But now she was oppressed by a sense of loneliness" (40). The isolation and distancing from other people and villages is pictured:

> Nagaveni stiffened. Previously there was no need for such questions. Where there was a village, there had to be people coming and going and talking. Not only the people who lived in it, had sometimes those who had to cross the Sharavathi gone past the house. Farm hands from Aralagodu too went by on their way to and from work. But now there was dread hovering about, a fear of strangers in the vicinity and questions like 'who?' 'Why?' entered the head.
>
> (41)

Nagaveni has a close attachment with a cow named Belli. Once, Belli did not turn up in the cattle shed. Nagaveni is anxious for the cow and starts searching for her. When Nagaveni calls the cow by her name, suddenly, "Belli, who had been grazing somewhere behind the carven, heard her mistress through the boulders, came and stood at the entrance and responded, '*ambaa*'" (24). This espouses Nagaveni's intimate bonding with the cow. Nagaveni's and Krishnayya's discussion is vividly contradictory and projects disparate perspectives about the heavy rain and its ramified consequence. Nagaveni's fear of the impending disaster is described, "I'm gripped with so many fears: What will happen when? What if water rushes into the house? What if the tiger or a boar comes in and takes away one of us?" In Krishnayya's response to her question one can see his patriarchal dominating attitude over animals, "Ayyo, you silly girl! With two men in the house, will we leave a woman to the tiger?' (55).

Starhawk, a spiritual ecofeminist, posits, "We see the earth as a living being, we are bound to notice what an erotic being she is. If you doubt me, go stand on a hill by the ocean with a spring wind blowing and a profusion of wild irises at your feet. Eat a strawberry, slowly. Or run your hand along the soft, bark of a redwood tree" (Foss et al. 1999:77). In the similar vein, this concept is projected through Nagaveni's character: "Nagaveni too looked delicate as if sprouting tender green leaves like everything else around them" (21). Nagaveni represents the union of women and nature through the different stages of nature. Krishnayya's words postulate his comparison of Nagaveni with rain, wind, clouds and sky: "Nagaveni too was like the rain, weeping for the past eight days; her face, the smoldering clouds, dark and angry; her sighs, the moaning wind from Sita Parvatha. . . . Nagaveni stay this way? Like low-lying dark clouds? Like raindrops blasted by the wind? Like the wind that whizzes down from the crouching boulders?" (75).

3.4.3 Nature as feminine sensibility

Most novels in the given study exhibit their ecofeminist concerns in delineating female characters. Such a conscious attempt is not made in *Dweepa*. On the contrary, the novel exhibits feminine sensibility through the portrayal of nature as woman. The feminist concern is portrayed through the changing moods of the river, the hillock, fields, rain and the landscape that culminate their own woes due to the introduction of developmental attitude in the form of dam construction. The river Sharavathi personifies the ethos of ecofeminism by its changing mood and draws an analogy between its course before and after the construction of dam. D'Souza postulates, "Now the time had come for the village to drown. Sharavathi had never come close to Hosamanehalli though she would roar ferociously from a distance during the monsoon. But she was now thinking of swallowing it up. These days her water did not flow freely; it stagnated in deep pools, waiting dangerously" (6).

Before the construction of the dam, the Sharavathi flows along by its natural course. The natural course of the river has never disturbed the crops of the village:

> But during the monsoon she was on the verge of overflowing. Even then, the water that overran the banks barely touched the paddy fields. That was about all. There was no danger to the fields or the farms. Now that the dam was being built across the river, the water level could surely rise higher. The government officials had installed a red stone on the forehead of Sita Parvatha behind the village to show how high the water would rise once the dam was ready.
>
> (3)

But, with the construction of the Linganamakki Dam, there are possibilities that the river may flow any side which is a peril for the villagers. Therefore, what will happen cannot be predicted? There are more possibilities of its engulfing the Sita Parvatha hillock and once the hillock is drowned then the entire area would submerge. The dam construction blocks the way of water by which the water is stagnated. Water, without rising, widens the riverbed on the side away from the hill and threatens to overflow the bank, which is very dangerous for Ganapayya's family to survive, "The Sharavathi lay like a pregnant woman, full and ready for birthing" (37).

Water enters into the fields of Ganapayya, which uproots a lime tree, a few banana plants. Next, areca plants are in danger of falling. Within a few days Hosamanehalli will be an island. While flowing, the Sharavathi River hugs the mountain. Its path is blocked in such a way that it begins to spread out by encroaching upon the neighbouring forest, the valley, the trees, the shrubs and water floods everywhere.

Next, the rain's changing mood shows the internal fury of the Ganapayya's family and Krishanayya – who comes from Nagaveni's mother's place – for their subsistence and survival. D'Souza says that the onslaught of dam construction on land reaches such a height by which "the land had become indifferent to the monsoon" (47) whereas Ganapayya's father oscillating between life and death metaphorically projects how the saplings of the fields are struggling to breathe: "The sapling in the fields stood breathing in the water as it flowed from field to field join the river. They were barely as tall as span as they shivered in the onslaught of wind and rain, yet they stood breathing in the slush" (47). The first monsoon has taken a toll on the palm trees and pepper creepers and all other trees; the only one thing that is rejuvenated is the pond because of gushing water.

The rainwater cascades from the Sita Parvatha to besiege the village Hosamanehalli by water as a small spot of island. Slowly, water engulfs Parameshwarayya's land and Heramba's house is also on the way to destruction. Next, water is waiting to enter Ganapayya's house – the only house that is still saved. Furthermore, due to the continuous heavy rain the wild animals enter Ganapayya's field which adds more woes to his worries:

> And now foxes, deer, and wild goats strolled fearlessly behind the house looking for shelter. A python crept into the wood-shack beside the kitchen. Rabbits scurried about the veranda. The cattle had mooed restlessly a few nights earlier. Wild animals like the tiger, cheetah, bison, and wild boar lived in the Malenadu forests but they lived in their own territory most of the time. Now with all the extra water around they could be scared too.
>
> (50–51)

D'Souza reclaims that the cause of the heavy rain is man's inhuman attitude towards nature. Usually *Aridharaa* rain pours mildly and *Marashir* rain downpours (these are the stars which influence the different phases of monsoon), but now the picture has changed. Even in *Aridharaa* rain starts pouring heavily. This is the consequence of man's indifferent action to nature: "Aridharaa seemed to be competing with Mriashira with a will to win" (53).

D'Souza has passed an ironical and wry comment on the human project of dam construction. Thus, we can sum up D'Souza's indignation towards development through the words of Shreedhara in the introduction:

> Of course, it has become a universal mantra of progress in contemporary Indian politics and is used as a cover up to usher in neo-liberal politics, structural adjustment programmes, and even communal violence. At the same time, it has led to various social movements and protests with issues ranging from environmental protection and livelihood questions to upholding human rights.
>
> (xix–xx)

3.4.4 Deep reverence for Indian agrarian life

D'Souza portrays very minutely the Indian men's deep reverence and attachment to their fields. Duggajja, the father of Ganapayya, has divine love to his field. Till his last breath, he is not ready to leave the land. When he comes to know that he has to leave the village Hosamanehalli, he feels as if his energy was sapped out: "He loved his piece of land with the attachment a woman feels for her mother's house. He was determined he would not leave her if he could help it. He had wondered a hundred times whether there was any way of carrying on here even when the village was covered with water" (12). His close bonding with his native abode and his love for the land are depicted through his desire to let the dam collapse so that there would be no need to leave their farm and village. He decides not to hurry because they have enough to eat. Though water may cover the village and even they won't get labourers to work, they would manage the planting and harvesting somehow. Thus, Duggajja and Ganapayya are against the modern developmental attitude which is in the form of dam construction.

3.4.5 Shortage of resources: threat to the livelihood

Ganapayya, the protagonist, is much worried about the shortage of resources which threaten their survival and livelihood. This is a consequence of human beings' greed and thoughtless attitude towards nature and non-human beings. The novel exhibits the paradoxical effect of dam construction on human beings and animals, which threatens the survival and livelihood of the native people. If we go on cutting trees and destroying the natural flora and fauna at this accelerating pace, then doom will not be far and the most important thing is there will be no one to utilise the fruits of development. The consequence of development is outlined through the words of Ganapayya:

> this year the monsoon was not as good as it used to be. . . . Malenadu was becoming barren with trees being cut to make way for railway tracks and highways telegraphy lines and dams and townships for outsiders. . . . if they continued to hack trees at this rate, of course, the rains will get scarce. And there won't be enough water in the Linganamakki Dam. The Sharavathi Project will be a waste.
>
> (76)

3.4.6 The uniqueness of Dweepa *as an ecohumanist narrative*

D'Souza's narration is not focused on the second key element of ecofeminism that is women as saviours and subjugated by the development attitude. His inclination is more towards men. The entire novel represents the agony of displacement which is shown through Ganapayya. Though the submersion of land affects both the man and the woman, D'Souza shows that compared to woman, man has more concern for nature.

This concept is shown through the protagonist, Ganapayya. He is very much attached to his land. However, owing to the dam construction on the river Sharavathi his land is going to submerge in water and he has to leave the land. Land is a means of their survival, "How can we live in this condition? I may say, enough of this problem, let's go elsewhere. But where can we go? I've depended on this farm and the field till now, where else can I live?" (11). Hermbha Hegde is allotted land as compensation near Ananthpura, so he has to leave the village. Ganapayya thinks to cultivate the land which Hegde leaves behind:

> Anyway, Herambha's leaving his lands. Why can't I harvest them with mine? He can't uproot his rice seedlings and areca palms; he can't take them with him. And he has no one here to watch over them. . . . I'll tell him I'll give him a part of the harvest as his share. This seems to be a good plan. If the officials do come and bother me to leave, I can always bribe them a bit. Let's see.
>
> (14)

3.5 Conclusion

The basic intention of this chapter is to examine how Indian novels portray environmental crisis and human displacement caused by the so-called developmental activity, the construction of dams. Markandaya tries to show how tribal people inevitably join the work of dam construction that indirectly affects their way of life. D'Souza's major concern is to draw our attention to the problem of submersion of land and the displacement of local people. In Markandaya's *The Coffer Dams* (2008), characters highlight the basic tropes of ecofeminism whereas D'Souza portrays the key elements of ecofeminism through conceptual thematic devices such as the dam construction and the changing moods of the Sharavathi river, the Sita Parvathi hillock, the rain and the flooded landscape. In addition, Markandaya and D'Souza bring to our notice that Indian environmental problems are necessarily human problems and not merely feministic issues.

Bibliography

Agarwal, Bina. "Environmental Management, Equity and Ecofeminism: Debating India's Experience." *Journal of Peasant Studies*, vol.25, no.4, 1998, pp. 55–95.

Cullet, Philippe. *The Sardar Sarover Dam Project: Selected Documents*. Ashgate, 2007.

D'Eaubonne, Francoise. "Feminism-Ecology: Revolution or Mutation?" *Ethics and the Environment*, vol.4, no.2, 1999a, pp. 175–177.

———. *Le feminism ou la mort*. Pierre Horay, 1974.

———. "What Could an Ecofeminist Society Be?" *Ethics and the Environment*, vol.4, no.2, 1999b, pp. 179–184.

D'Souza, Na. *Dweepa*. Translated by Susheela Punitha. Oxford University Press, 2013.

Eaton, Heather, and Lois Ann Lorentzen, editors. *Ecofeminism and Globalisation*. Rowman & Littlefield Press, 2003.

Foss, Karen, et al., editors. *Readings in Feminist Rhetorical*. Sage Publications, 1999.

Guha, Ramchandra. *An Anthology among the Marxist and Other Essays*. Permanent Black, 2006.

———. "Prime Minister and Big Dams." *The Hindu*, 18 December, 2005.

Markandaya, Kamala. *The Coffer Dams*. Penguin Books, 2008.

Mellor, Mary. *Breaking the Boundaries: Towards a Feminist Green Socialism*. Virago, 1992.

———. "Feminism and Environmental Ethics: A Materialist Perspective." *Ethics and the Environment*, vol.5, no.1, 2000, pp. 107–123.

Merchant, Carolyn. *The Death of Nature*. HarperCollins Publishers, 1980.

———. *Earthcare: Women and the Environment*. Routledge, 1995.

Mies, Maria, and Vandana Shiva. *Ecofeminism*. Rawat Publications, 1993.

Murphy, Patrick D. *Ecocritical Explorations in Literary and Cultural Studies Fences, Boundaries and Fields*. Lexington Books, 2009.

———. "Ecofeminism and Postmodernism: Agency, Transformation, and Future Possibilities." *Women, Ecology, and the Environment*, vol.9, no.3, 1997, pp. 41–59.

Rao, A.V. Krishna. *The Indo-Anglian Novel and the Changing Tradition*. Rao & Raghavan, 1972.

Scheumann, Waltina, and Oliver Hensengerth. *Evolution of Dams Policies*. Springer, 2014.

Sharma, Lakshmi Kumari. *The Position of Women in Kamala Markandaya's Novels*. Prestige Books, 2001.

Shiva, Vandana. *Biopiracy*. Natraj Publishers, 2012.

———. *Staying Alive Women, Ecology and Survival in India*. Women Unlimited, 2010.

———. *The Violence of the Green Revolution*. Zed Books Publication, 1993.

4 The industrial disaster

Animal's People

4.1 Introduction

The major ecofeminist contention is that the idea of the modern development is the root cause of exploitation of women and nature. Vandana Shiva, a prominent Indian ecofeminist thinker and activist, attacks the idea of development as a Western paradigm and owing to its adverse effects on human beings in general, and women and nature in particular which she considers as mal-development (Mies & Shiva 1993; Shiva 2010). The basic purpose of development or progress is economic gain, which is based on Western industrialised economy. The terms 'development,' 'progress' and 'globalisation' resonate with the values of Western culture which focuses on productivity and economic growth rather than concern for non-human beings. The arrival of the Western industrial development has been the major source of ecological crisis. In the present chapter, we will examine Indra Sinha's *Animal's People* (2007), which represents the horrendous impact of an industrial disaster on human beings and nature.

Indra Sinha, the winner of Commonwealth Writer's Prize, has translated many Sanskrit works into English, though his mainstay is science fiction. His major works include translations of ancient Sanskrit scripts into English – *The Love Teachings of Kama Sutra* (1980), *Tantra: The Search for Ecstasy* (1993), *The Great Book of Tantra: Translations and Images from the Classic Indian Text* (1993). He has written non-fiction including *The Cybergypsies: A True Tale of Lust, War, and Betrayal on the Electronic Frontier* (1999) and one more novel *The Death of Mr Love* (2002).

Indra Sinha's *Animal's People* (2007) is an attempt to gauge the worst effects of Bhopal gas tragedy on human beings and nature through a dystopia, Khaufpur. Sinha narrates the novel in a unique way. Animal is a name of the teenager boy in the novel. The impact of the disaster bends him at the bottom of his spine, which forces him to walk on all fours just like an animal. This identity is the gift of the industrial disaster. He is an orphan and brought up by a French nun Ma Franci. This novel is a saga of the pathetic condition of the boy and his life, which represents thousands of Bhopal youngsters' lives. Further, it is an account of the sufferings of the survivors of the impact of the industrial disaster. On parallel lines, Sinha shows the impact of the disaster on a lake, animals, water sources and nature. The narration of the novel is through the central character – Animal. He recounts the story to a tape recorder for his journalist friend who is going to publish that recording as a book. Each vignette is named tape number one, two and so on.

4.2 The Bhopal gas tragedy: a backdrop

The novel is an outcome of Sinha's sheer personal experience of working for the victims of the Bhopal gas tragedy, which has been the world's worst industrial disaster so far. Indra Sinha attempts to gauge the deadly disaster out of his personal ethos and for whom he worked to collect funds. In the year 1984, the pesticide factory of Bhopal, owned by the multinational company Union Carbide, accidentally released deadly toxic gas that took a heavy toll on the lives of the people. This world's worst industrial disaster released deadly toxic gases – a phosgene and a toxic concoction, methyl isocyanate gas (MIC). These gases were accidently released from the factory. They contain a cyanide base, which causes many deadly diseases:

> Your eyes, throat and lungs begin to burn and fill up with oozing fluid and melting tissues. Blinded, you gasp for breath as fluid begins filling up your lungs. Then you lose control of your nervous system, you vomit uncontrollably, cramps seize your stomach. If you are lucky, you lose consciousness quickly and you die. If you are not, your death is a long drawn out, agonizing affair. If you survive, your lungs and eyes will never work properly again. Muscle pains and ulcers will prevent you from working or leading a normal life.
>
> (Mukherjee 2010:135)

In this incident, many were injured with no hope of recovery. There was a massive damage to plant and animal life. And in the aftermath of the disaster, there were long-lasting effects on the survivors. This industrial disaster affected 200,000 people out of 900,000 people living in the surroundings. Almost 10,000 people lost their lives. And 60,000 are suffering from injuries; they are almost nearing their death (Mukherjee 2010:134). Sinha projects his reflection of this incident through his literary endeavour.

4.3 Patriarchal developmental attitude: industrial disaster

We scrutinise the impact of the industrial disaster on human beings and nature at two levels. One, the US multinational company Union Carbide repudiates its responsibility to pay proper compensation and to clean the poison and deadly residue of the disaster. Two, we explore the massive impact of the industrial disaster on survivors and nature. First, the US multinational company built a pesticide factory in a highly populated urban area. "Why did the Kampani choose this city to make its factory? Why this land?" (32). Sinha documents the inhumane attitude in the form of development through this novel: "It raises the question of the politics of environmental toxicity – who decides to build (or dump) what, where and how – which affects a disproportionate number of human and non-human beings who have little say in the matter. The tragedy raises questions about international frameworks of law, justice and rights, or more appropriately here, the lack thereof" (Mukherjee 2010:134).

Second, the toxic gas released accidentally into the air resulted in the deaths of thousands of people that night and many more were the worst survivors and first-hand victims of the disaster. The survivors are still facing manifold consequences through their physical and mental disabilities. We can have a glimpse of a few – Animal, the protagonist and the narrator of the novel, walks on all fours; Aliya, the young girl, coughs throughout the day and meets her end; Ma Franci, the nun, has lost her ability to recognise languages and in her later days talks deliriously; in every house of Khaufpuri, more than one victim struggles hard for survival.

The US based owners washed off their hands from their responsibilities. First, they run away without cleaning the factory. The poison is left behind in a massive amount, which is found in the wells of Khaufpuri. Everyone you meet seems to be sick: "The factory is abandoned full of chemicals which as we speak are poisoning the water of thousands more. Must all perish before these Amerikan defendants appear?" (Sinha 2007:52) Second, they are not ready to pay proper compensation to the survivors. For more than 18 years the court has not given any verdict to the victims of the industrial disaster because the American defendants are not ready to turn up at the court: "They have not even bothered to send lawyers. They sit in Amerika claiming this court has no jurisdiction over them . . . thus these proceedings drag on and on, for the people of this city justice continues to be delayed and denied" (52). Yet, the Khaufpurians are waiting for the day on which the site of the factory will be cleaned and they get proper compensation.

Third, the company owners have been trying every possible way to get rid of the cleaning of the factory and compensation. For this many lawyers are appointed, many doctors have been doing research to show that the accident has not caused much harm to the people, and many Public Relations Consultants are appointed to deal with protesters like Zafar. Furthermore, to have some control over the victims of Khaufpuri, they throw parties for the elite, like "judges, senators, presidents, oil sheikhs, newspaper owners, movie stars, police chiefs, mafia dons, members of royal families, and prime ministers." Zahreel Khan, Khaufpur's a Minister of Poison Relief, promises to wipe the tears of the poison victims. In real terms, nothing is worked out; every hearing of the court is postponed. Zahreel Khan, during his visit to the place Nutcracker in front of a crowd of journalists, drinks a glass of well water to show that the well water in that place is safe. But, a minute later he goes behind the house and spits out water. Along with him the Chief Minister and many more ministers are deaf to the plea of people. Indifference of the government and the factory owner (USA) towards their responsibility of cleaning the site and giving proper compensation to survivors is shown through the plight and suffering of the people of Khaufpur.

Sinha exhibits the impact of the industrial disaster on people and nature at various levels. First, we map the protagonist Animal's physical deformity and mental trauma. Second, a few families and their agonies are projected. Subsequently, we espouse the impact on birds, animals, water and the natural phenomena. Residues of the disaster are long lasting. They contaminate water and pollute air and land to such an extent that it is high time to come out of the deadly disaster. Eventually, we notice the unique technique of narration; for example, Sinha compares the day-to-day activities of human beings with the natural activities through the use of rhetorical devices to create awareness among the readers that nature plays a very vital role in our lives.

To emphasise the horrendous impact of the industrial disaster, first, we examine the character of Animal – the epitome of the young generation. The very opening lines of the novel give us the idea of the juxtaposition of life before and after the industrial disaster: "So sweet you were, a naughty little angel. You'd stand up on tiptoe, Animal my son, and hunt in the cupboard for food" (1). This character is examined at three levels. First, his physical deformity is a result of the disaster, which has taken a toll on his identity: Though he is a human being, he is named Animal. The entire novel narrates the juxtaposition of two identities: before the disaster, he was a human being and was leading a very normal and happy life. But after the disaster, he is similar to animals not only in physique but also in temperament. Second, his physical deformity impairs his mental growth. Third, he hates to be a human being. Fourth, Animal narrates the long-lasting effects of the disaster on people. Even, Khaufpurians don't want to recall the horrifying incident.

Animal is reared by the nuns, especially, by Ma Franci: "On the night I was found lying in a doorway, child of a few days, wrapped in a shawl. Whose was I am? Nobody knew. Mother, father, neighbors, all must have died for no living soul came to claim me, who was coughing, frothing etc plus nearly blind, where my eyes had screwed themselves against the burning fog were white slits bleached on the eye-balls" (14). After the disaster, he gets a burning sensation in his neck and shoulders. He is not able to lift his hand. The pain is very severe due to which he catches fever. He is hospitalised where he is given an injection, but his pain is not reduced. The intensity of the pain is aggravated by which he couldn't straighten up his body, "When the smelting in my spine stopped the bones had twisted like hairpin, the highest part of me was my arse" (15). His back begins to twist and he is not able to stand straight.

He bends forward and the upper part of his body is strong, but the lower part is deformed. Therefore, he has to walk on hands. The boy's mental agony is mapped through his casual way of the narration of people's suffering, which gradually acquaints us with his new way of life, which has turned out to be a curse for him.

The way he walks on his hands makes the orphanage children call him 'Animal.' In addition, his behaviour has named him Animal. He is tough enough to participate in the game called kabbadi. He is even good at catching the opponent player; but we can see how bitterly he resents and is disgusted with his peer group: "One day I grabbed this boy, he kneed me in the face. It hurt. I was so angry I bit him. I fastened my teeth in his leg and bit till I could taste blood. How he yelled, he was howling with pain, he was pleading, I wouldn't stop. I bit harder. The other kids started shouting, 'Jaanvar, jungle Jaanvar.' Animal, wild animal" (15). This incident shows two things: his hatred towards the other normal children and his inhuman activity that makes children call him animal.

The other incident to name him Animal is: one day he is lying on the grass on the bank of the lake because he is not able to swim. In the meantime, a girl of his age pushes him and puts her muddy fingers on his body and says, " 'Like a leopard!' So then they all dipped their fingers in the clay and covered me with leopard marks. 'Animal, jungle Animal!' The name, like the mud stuck. The nuns tried to stop it but some things have a logic that can't be denied" (16). This is the way the young boy gets his identity changed, which leaves its indelible marks on his mind in such a way that he undergoes a mental trauma.

Though Sinha depicts Animal's physical deformity, the core intention is to show its impact on his psychology. His disgust and resentment vent out through his hostility towards mankind: "I hated watching my friends play hopscotch. I detested the sight of dancers, performing bears brought by those dirty buggers from Agra, stilt-walkers, the one-leg-and-church of Abdul Saliq the Pir Gate beggar. I envied herons, goalposts, ladders leaning on walls. I eyed Farouq's bicycle and wondered if it too deserved a place in my list of hates" (2). Moreover, he is not fit to do any work: "I don't know. No one has ever trusted me with anything before" (25).

Next, we will focus on how the novel deploys contrasting identities of Animal. Before the disaster his life is very much normal and very happy, in the words of Ma Franci, "So sweet you were, a naughty little angel. You'd stand up on tiptoe, Animal my son, and hunt in the cupboard for food" (1). He recalls his blurred childhood memories narrated by Ma Franci that he was a handsome boy with huge eyes and a beautiful smile. He could walk upright on his feet like a human being. Even, he is able to swim in the lakes: " 'You'd dive right in, with your arms and your legs stretched out in one line.' Whenever she said this I'd feel sad also angry. I still dream of diving straight as a stick into deep water leaving my crooked shadow behind" (14).

On the contrary, everything shatters in a blow of a whirlwind in the form of the explosion. The post effect of disaster on his mind and body can be traced to his words: "My story has to start with that night, I don't remember anything about it . . . time divides into before and after the before time breaks into dreams, the dreams dissolve to darkness" (14). The novel features the mindscape of Animal, which leads to hatred first for the mankind and second for himself to be a human being. Even, he doesn't want to look into a mirror even by chance, as he hates his physical deformity. He hates not only himself as a human being but also the entire mankind: "I'd be filled with rage against all things that go or even stand on two legs. The list of my jealousies was endless" (2). He hates his friends playing hopscotch and dancing; even he hates herons, goalposts, ladders and bicycles, all of which can stand erect. The reason behind this animosity is that the world of human beings is viewed from the eye level: "Believe me, I know which one hasn't washed his balls, I can smell pissy gussets and shitty backsides whose faint stenches don't carry to your nose, farts smell extra bad" (2). From this perspective, Animal has developed a sense of absolute difference from those human beings who are powerful enough to possess abstractions such as 'rights' and 'justice.'

He has revulsion against human beings because they fail to understand his agonies and problems. Instead, they make fun of him. They harass him. By pointing at him, they say, "See, that's him, bent double by his own bitterness.' People see the outside, but it's inside where the real things happen, no one looks in there, maybe they don't dare" (11). Even it is a puzzle for people how he manages to shit: "somethings have logic that can't be denied. How do you shit, when your your arse is up in the air and legs too weak to squat? Not easy. What do you look like as the turds tumble from your hindquarters? Like a donkey dropping dung" (16). The comments of his mates' mark a deep scar on his heart: "In my streets years I hated to see dogs fucking, my mates would shout, 'Hey Animal, is this how you do it?'" (16). He faces humiliation at every level and many times loses his temper.

Being a human being, he is not able to fulfil his basic instincts. An incident can be quoted from this point of view. After learning how to read and write from Nisha, Animal demonstrates his fluent reading. Nisha's happiness knows no bounds and she hugs him. His expressions and feelings are mapped through his remarks: "No girl had ever touched me till then, less hugged. It sent a thrill through me straight to my cock. This was the first time I caught myself thinking, if only things were different with me, if I could walk upright, it might be my praises she sang instead of Zafar's" (36). Here, he thinks that if he was normal like all other human beings, he might have fulfilled his sexual instinct.

People call him *jaanvar*. The word *jaanvar* denotes that *jaan* means 'life' and *nvar* means 'one who lives.' This elucidates that the one who has life, either a human being or an animal, makes no difference. In this context, one could quote Mukherjee:

> They ensure that it is no longer possible to see 'nature' or 'environment' as something that exists out there, outside the realm of the human or the social. Rather, environment must be seen as a mutually sustaining network in which humans and non-humans are always already linked with each other, and on whose collective action and prosperity the functioning of the network depends.
>
> (Mukherjee 2010:47)

Though Animal is a human being, he perceives himself as an animal. He says, "My name is Animal. I'm not a fucking human being. I've no wish to be one" (Sinha 2007:23). Animal's reluctance to accept the identity, rights and privileges of human beings shows the condition of non-human beings. He always dreams that he will go to New York for surgery so that he can walk upright and he can fulfil his carnal desire. But, he does not get the rights of human beings and so, asserts, "I am not a man. I am filled with revulsion for human life and human society, I want no more of it" (303 & 341).

Let us study another character, a victim of the industrial disaster, a dog named 'Jara.' She is a co-mate of Animal in scavenging for food on the streets. They are friends as well as enemies and rivals for food. Sinha's narrative structure is highly focused on equality of man and animal as well as highlights how both are the congruent victims of toxic gas. They roam for food late in the nights. They wait for waiters to throw scraps of food at street corners. Jara and Animal fight for food – for a bit of naan, banana skins and a few shreds of mutton. This pathetic condition of Animal as a human being and dog as an animal is ubiquitous in the city of Kaufpur. This parallel narration spotlights a palpable synergy between human beings and animals:

> The fish was too good to give up. I stuck firm as she made her approach, the lips lifted over those evil yellow teeth. She started all that rrrrr business. I don't know how, but some rebellion ignited inside me. On all fours I rushed at her snapping my jaws, growling louder than she, the warning of desperate animal that will stick at nothing. She turned and slunk back a few paces, then lay down again, giving me a reproachful look.
>
> (17)

This example shows Animal's behaviour just like a dog. Further, Animals' pathetic condition is portrayed through Jara. Her physical condition projects her hunger and starvation: "She was as thin as me, her hide shrunken over her ribs" (17). This represents how both Jara and Animal are the consequences of the disaster. Jara has a sore and leaking nose. Animal says, "I named her Banjara, gypsy, free spirit, because she belongs to nowhere and everywhere is her kingdom" (18). When Animal starves for food, he sells his blood for Rs 80. He also scavenges for rags, tin cans and plastic to fulfil his hunger. He is not able to do any other work because he has to go on all fours. He has only one hand plus his mouth free to carry things. The core objective of Sinha is to sketch these two characters – Animal and Jara – at par to represent human beings' and animals' desperate condition because of the industrial disaster.

Other children Animal's age go to school to get an education in reading, writing and math; but poverty, hunger and starvation have schooled Animal in haunting and stealing food along with Jara. Animal considers the natural instinct as unnatural because mentally he determines that he is an animal:

> Animal mating with human female, it's unnatural, but I've no choice but to be unnatural. Many times I would dream that she and I were in love, sometimes we were married and naked together like in the movies having

> sex. In such dreams was my back straight? Did I stand upright? No and no. I was exactly as I am now and it did not matter. Such dreams! I woke from them shaking with hope.
>
> (78)

Let us probe into the next character Somraj, father of Nisha, who is a singer. He is known throughout India as Aawaaz-e-Khaufpur, the voice of Khaufpur. He sings on radio and gives concerts. However, the horrible night has taken away his wife, baby son and even damaged his lungs. The disaster not only grabs his breath but even his life:

> it also stole his life, because breath is the life of a singer. From that night on he would listen to other people's records, but never his own. He became a solemn and private man. Later he started teaching music, his students won prizes, to them he was like god but he seemed to get no pleasure from it since hardly ever would you see him smile.
>
> (33)

The impact of the disaster is traced to the Animal's friend Faqri, who loses his mum, dad and five brothers and sisters. Along with them many people died in the lanes: "Mira Colony, then Khabarkhana and Salimganj. East's Phuta Maqbara, to the West Qazi Camp, killing grounds all. On that night it was a river of people, some in their underwear, others in nothing at all, they were staggering like it was the end of some big race, falling down not getting up again, at Rani Hira Pati ka Mahal, the road was covered with dead bodies" (31–32). On that night almost all people lost their families, friends, health, jobs and everything.

Next, the disaster badly hits the orphanage in Jyothinagar. Many children die and even nuns too. Those who survived are the sick. Huriya, the old grandfather of Alyia, is a victim of the disaster. On the one hand, he suffers from an ailment and on the other hand he lost his daughter and the only ray of hope in his life is his granddaughter Aliya, who suffers from lung infection. Shambu is a double victim of the disaster: "He had breathed the poisons of that night, plus the wells in his neighbourhood were full of poisons leaked from the factory" (147). He is hardly able to breathe and Yusuf Omar is caught with ulcer. The nightmarish impact on the heath of the people is traced to the Elli doctors' way of treatment: "the Kampani's that-night-poisons have damaged people, such as eyes, lungs, joints, womb, brain. These are marked in red. In blue are marked the places which have been harmed by drinking the poisoned water, breasts, again womb, stomach, skin. Blue and red spirals are coming from the head" (162).

In the aftermath of the disaster, the effects are manifold. First, they lose their means of survival. Second, they are not able to work because of deterioration of their health, which leads to hunger, starvation and frustration that "there is no night and day, only a vast hunger through which suns wheel, and moons wane and wax and have no meaning" (185). Many times, they do not get a square meal a day. Therefore, they have to live without food:

> When you can't remember the last night you ate something? I'll tell you. When it's light there's binding a cloth tight round your belly to squeeze out the pain. When it turns dark you've to drink plenty of water to fill your miserable gut. Hope dies in places like this, because hope lives in the future and there's no future here, how can you think about tomorrow when all your strength is used up trying to get through today? Zafar says this is why the people don't rise up and rebel.
>
> (185)

Let us explore the impact of the disaster on nature and animals. Owing to the release of the deadly gas, everything is dead; nothing is left alive, "No bird song. No hoppers in the grass. No bee hum. Insects can't survive here. Wonderful poisons the Kampani made, so good it's impossible to get rid of them. . . . peacocks, goats and even grey herons were found dead beside the kampani's lakes" (29 & 49).

Sinha projects the cold war between the natural and the artificial world. Earlier, the land was full of vegetation, but this natural life is replaced by the construction of the factory. But Mother Nature tries to get back her land by growing wild sandalwood trees, "Under the poison-house trees are growing up through the pipework. Creepers, brown and thick as my wrist, have climbed all the way to the top; tightly they've wrapped wooden knuckles round pipes and ladders, like they want to tip down everything the Kampani made" (31). Further, water in Khaufpuri is contaminated to such an extent that almost all the people are sick. The sarcastic comment of Animal very clearly projects, "although Khaufpuri pump water stinks it is free" (102). Elli, the doctor, appeals to her husband Frank to help her save people from contamination of water. She explains, "Its poisons are in the wells, they're in people's blood, they're in mother's milk. Frank, if you came to my clinic I could show you. Specimens, I mean. Foetuses, babies that never made it. You wouldn't want to see such things, even in your nightmares. . . . what ways specially is the water affecting people's health? What kinds of illness are showing up?" (322). Till the leakage of the toxic gas, the place is Paradise Alley; but the worst night has taken a heavy toll, "Paradise Alley is wreckage of baked earth mounds and piles of planks on which hang gunny sack plastic sheets, dried palm leaves" (106). Once the lake water was pure but due to the toxic gas, the "lake looks pale green from up here, eye slides along a road lined with dirty buildings" (133). Next, the impact of the industrial disaster is on ponds and gardens, which have dried up, and the consequence is that fish live only in plastic tubs.

4.4 Women as victims of the industrial disaster

Major ecofeminists think that women are first-hand victims of the catastrophic effect of the industrial disaster. This concept is well represented through a few characters: Ma Franci, Nisha, Aliya, Huryia, Pyrebai, Lilabai, Sahara, Nasifa,

Safiya, Anjali and so on. Ma Franci is a French nun who works in the orphanage. She has reared and taken care of Animal as her own child. She lost her ability to recognise languages especially Indian, among them is Hindi, which is her base to be a teacher to teach in the school and communicate with Indian people especially Khaufpuri people:

> Ma Franci lost knowledge of Hindi. She'd gone to sleep knowing it as well as any Khaufpuri, but was woken in the middle of the night by a wind full of poison and prophesying angels. . . . She forgot all languages except her childhood speech of France. . . . But there was a further twist to Ma Franci's madness, when she heard people talking in Hindi or Inglis, or come to that in Urdu, Tamil, Oriya, or any other tongue used in Khaufpur, she could no longer recognize that what they were speaking was a language, she thought they were just making stupid grunts and sounds.
>
> (37)

Many French nuns are sick and so they are sent back to France. But Ma Franci refuses to leave. She came to India 40 years ago and dedicated her life to serving the poor and the orphans. Nisha – a daughter of Somraj, a 20-year old Hindu girl – joins Zafar in his project of fighting against the victims of the industrial disaster. She loses her mother and brother. Moreover, though her father is alive; he has lost his life. His life is his voice; that is, singing classical music was his life but his lungs are affected by the toxic gas and he is not able to sing anymore. Aliya is a young girl. Her lungs are inflamed and infected. She coughs and has a high fever, and many times she has gone into an unconscious state. Finally, she meets her death at a young age. Lilabai needs blood for operation, but it is too expensive to buy. Many women, who are expecting that night, miscarry. Pyre Bai is the wife of a factory worker, Aftaab. He is a victim of the toxic release. He loses his job. He is not able to breathe properly and so, not able to do any physical work:

> When her husband got really ill and could no longer work, they ran out of money and had to sell their small house. They moved to a rented place with half a roof. It was the only place Pyare could find, right by the stinking naala, in the monsoon the rain came right in. The small girls were always hungry. At night they cried. She would bind cloths round their waists and give them water to fill their empty bellies. She found a job to carrying cement on a building site.
>
> (84)

She loses her husband in spite of all her efforts. Moneylenders take away her cooking pots and her husband's bicycle. Another heart-rending suffering is of a mother who's scared to nurse her child because she thinks her milk has toxic content so she spurts her milk on the ground. "'I won't feed my kid poison' she's leant forward to cast the last dribbles of her milk onto the ground" (107).

Sahara, a 46-year-old woman, develops cancer. Her womb starts bleeding and she meets her end. Nasifa, another woman, has a swelling in her neck. As a result, she is not able to lift her arm as "it felt like someone pulling her nerves from the inside" (109). She is not able to survive. Safiya, a pregnant woman, has pain in her womb. She always feels that any moment she is going to miscarry. The doctor suggests her to drink milk and eat fruits. She goes to a shopkeeper and asks him, "Zinda, you must help us. We can't even afford rotis, how will we afford fruit?" "I gave her some guavas, I said," "Sister, pay me when you can. But before she could pay, she was gone" (109).

Hanif, the grandmother of Aliya, is left blind and has been coughing for the last 20 years from the day of the disaster. Anjali, one who works on the fields, turns to the profession of prostitution. Devika and one more lady from the place called Blue Moon colony are the victims of the deadly disaster. Another lady is Gargi, whose back is bent like that of Animal. When she comes to know that she will not get proper justice, she says to her lawyer, "Mr Lawyer, we lived in the shadow of your factory, you told us you were making medicine for the fields. You were making poisons to kill insects, but you killed us instead. I would like to ask, was there ever much difference, to you?" (306). Many girls menstruate three times a month and some have once in five months. Doctors do not know how to treat them. These are a few illustrations of women are first-hand victims of disaster.

Thus, the industrial disaster induced by man's selfish motive leads many people to lose their lives and many live as cursed beings. This can be witnessed in Animal's words, "I would like at the lights of the city and wonder if this pipe had been mended, that wheel tightened, I might have had a mother and father, I might still be a human being" (32).

4.5 Rhetorical tropes

Indra Sinha uses a unique technique to delineate the interrelationship between man and nature. He compares man's day-to-day activities with that of animals, nature or natural phenomena. Animal says that birds represent our words, "Blue kingfishers won't suddenly fly out of my mouth" (2). The boys' quivering quality is shown through, "This boy stopped talking, queer as a winged snake" (4–5). Human beings' inhumane attitude is represented through, "Like vultures are you jarnaliss. Somewhere a bad thing happens, tears like rain in the wind, and look, here you come" (5) and "The things squeaked like a rat having its back broken and I heard my own voice earning fifty rupees" (6). Human beings' intimacy and his ability to understand the feelings of birds and animals are very aptly exposed in Animal's words: "I could see their souls. Most were ugly; some shone like green birds, but all without exception were full of fear" (11).

Sinha espouses two things: first, he compares our way of speaking with the birds' action of flying. The second, the minute picturisation of the process of flying of the birds captured in these words: "I can feel it coming, words want to fly out from between my teeth like a flock of birds making a break for it. You know sudden clap of wings when they take off in hurry" (12). Many human actions are compared with birds, animals, natural course of action, "Like water flooding into a field, the new language just came" (35–36). Here, our thought process is compared with birds' sounds: "In my head a thing flees away shrieking like a bird, eee-chip-chip-chip, the sound of the world dwindles to an eeire hum" (57). The action of speaking in a difficult situation is just like chirping of cricket: "I'm trying to speak, but my voice is a chirping cricket that hops from my tongue and is lost" (342). The cobra's action is shown: "My mouth opens a cobra up out of my throat its body fills my guts its tail dangles out of my arsehole every muscle in my body strives to expel it, up comes nothing" (344).

The fundamental point behind analysing a few sentences of the novel is to show Sinha's intimate association with nature and his in-depth knowledge of birds, animals and non-human beings. We can delve deep into the minute and detailed comparison of human beings' actions either with birds, animals or with natural phenomena to put forth how nature is an integral part of human beings.

4.6 The uniqueness of *Animal's People* as an ecohumanist narrative

Compared to all other Indian novelists' presentations of ecofeminist concern, Sinha traces the elements of ecofeminism differently: he focuses on the modern developmental attitude and its impact on both women and men. Therefore, we can say that Indian environmental problems are human problems. We traced a few ecofeminist elements in novel; however, the novel has not directly displayed that men are the saviours of the nature. But the thematic analysis of the novel provides a perceptive insight into the ecofeminist discourse. It shows man's concern to nature through his concern to human beings. In spite of engaging with the theoretical underpinning of the ecofeminist discourse, Sinha encapsulates that men are also saviours of human beings and nature through certain characters. We can infer a constructive underlying meaning of the conceptual understanding of the ecofeminist discourse through the impact of modern developmental attitude on women and nature.

On the other hand, man's concern for nature is shown through their contention of industrialisation and modern developmental attitude. They fight against the company to get the justice for the people. It exhibits that they are not ready to accept the modern developmental attitude in the form of industrialisation, which is the root cause of the exploitation of women and nature. Therefore, there are a few characters such as Zafar, Somraj and Farooq who put forth that men are also saviours of nature.

Zafar, a legend in the Khaufpur, has given up everything in his life for the sake of the poor and the victims of the industrial disaster. He lives his life like one among them and drinks water from the same stinking well. He counsels Animal on the physical deformity: "You should not think of that way, but as especially abled" (23). He is the most brilliant student of the college. He could have reached any position, but when he gets the news of the dark night of the industrial disaster, he quits college and goes to Khaufpur to organise the fight against the company, "Who do you suppose has kept the case against the Kampani alive so long? So many times they try to stop him. He's been threatened, beaten, but Zafar is not afraid of anything or anyone. He speaks the truth and he never gives up. These are the things that make the ordinary people love him" (27). He has been helping almost all needy people. To some he gives roofs: "Careful with this, it's for a family who need a roof on their place. . . . It's eleven and a half thousand rupees" (39). For Lilabai, he provides money to buy blood: "Animal, hardly can people afford food, how can they spend on medicines? This is why we have to help. Yesterday I had a case, a woman Lilabai needed blood for an operation" (24). When he hears Pyre Bai's pathetic story, his heart stirs and he says that "but I know that if I do not help that poor woman and her daughter, I myself will die" (85–86). Even to help poor people, he wants to raise funds and after payment of debt to Pyre Bai, he thinks, "we'll begin new work, a trust to give low interest loans so people are not driven to these scum" (86). He is ready to dedicate his life for the cause of providing justice to the people of Khaufpuri.

Zafar and Farooq have gone on a hunger strike in order to give justice to the people, at considerable risk of their lives, "Everyone knows that if tomorrow comes and if Zafar and Farouq do not call a halt, by day's end they will surely die. Eerie is the silence of so many people, almost, you can hear them breathe. Farouq and Zafar are lying on rugs with their eyes closed, a few people are sitting silently by" (299). Zafar has a lot of concern for people, animals and nature. The sufferings of the people of Khaufpur are out of control. He says, "We will take our whole crowd to somewhere outside the city, some spot where there's trees plus water, we'll take with us bread and chicken and sweets, we'll make tea on a fire of sticks" (54). Zafar pours out his heart to help the needy people. He remarks:

> The Kampani must return to Khaufpur, remove the poisons from its factory plus clean the soil and the water it has contaminated, it must pay for good medical treatment for the thousands of people whose health it has ruined, it must give better than one-cup-chai-per-day compensation, plus the Kampani bosses must come to Khaufpur and face the charges from which they have been running for so long and the court case against them should conclude.
>
> (227)

Somraj, a well-known musician, runs a poison-relief committee to help the local people who are not getting proper treatment, who are coughing up their lungs out for so many years after the disaster and those people who are least cared by politicians and people who find no lawyers to take up their claims for compensation. He actively participates in all the protests to fight against the US based owner.

The major first-hand victim of the industrial disaster is the protagonist – Animal. After the disaster, he gets a burning sensation in his neck and shoulders by which his back begins to twist and he is not able to stand straight. He exasperatedly says, "On my hands I learned to walk, my legs grew feeble. My hands and arms are strong, my chest is strong. The upper half of my body is like a bodybuilder's. I walk, also run, by throwing my weight onto my hands, hauling feet forward in a kind of hop. It took a long time to master this new way of getting about. Maybe it was months, maybe a year" (15). Though his half body part is strong, he is not able to work because he has to walk on all fours. His physical deformity affects his psychology. He hates himself and human beings. He never thinks of himself as a part of human society. He has no identity. He is neither able to be a human being nor an animal. Next, the victim of the disaster is Somraj, the father of Nisha, who has lost his wife and son along with his voice. As a singer, his voice is his life.

Further, Animal's friend Faqri loses his mum, dad, five brothers and sisters. Aliya's grandfather, Hanif, is blinded and continuously coughing; Shambhu can hardly breathe; Yusuf Omar catches an ulcer. Also, Rafi, being seriously ill, spends all his money on medicines. He has hardly any money to spend on food. But the medicine proves to be of no use; he dies. Aftaab works in a factory. He knows the chemicals produced in the factory are very dangerous. When the dangerous toxic gas releases into the air, he covers his two daughters' and Pyaré Bai's faces with wet cloths and carries them to a distant place by which he is able to save them, whereas his neighbours perish. Yet does not take any care of himself. Therefore, he continuously coughs foam tinged with blood. He has rashes all over his body and suffers from fever and has pain in the joints. Pyaré Bai says, "When I started work, my husband apologized to me for pitting me through all this. How often did he tell me not to spend money on him and his illness? Don't waste your money, he said, I'm going to die anyway. And he did . . . he left me alone" (84–85).

4.7 Conclusion

Indra Sinha's *Animal's People* demonstrates that man's developmental attitude has taken a heavy toll on the lives of human beings and nature. The novel graphically showcases the impact of industrial disaster on human beings and nature. Like other novels discussed in the previous chapters, *Animal's People* also depicts men and women as saviours who are subjugated by the modern developmental attitude. Therefore, Indian environmental problems are not merely preoccupied with gender issues.

Bibliography

Mies, Maria, and Vandana Shiva. *Ecofeminism*. Rawat Publications, 1993.

Mukherjee, Upamanyu Pablo. *Postcolonial Environments: Nature, Culture and the Contemporary Indian English Novels*. Palgrave Macmillan, 2010.

Shiva, Vandana. *Staying Alive Women, Ecology and Survival in India*. Women Unlimited, 2010.

———. *The Violence of the Green Revolution*. Zed Books Publication, 1993.

Sinha, Indra. *Animal's People*. Simon & Schuster, 2007.

5 Animals as absent referents

The Man from Chinnamasta

5.1 Introduction

The central concern of ecofeminism for nature is not based on the philosophy of 'nature for the sake of nature,' but since the health of the natural world plays an important role in keeping the health of human beings, nature is taken into account. Though the basic focus of ecofeminism is to resolve the problems of women and nature, a decade ago there was a critical development in this field. Animal Studies and Naturalised Epistemology added a new perspective to ecological crisis by theorising interspecies religious. The animal studies addresses interspecies or gender justice called animal ecofeminism (Gaard 2011). The consideration of the animal world is an important development in the domain of ecofeminism. The major ecofeminists comprehend that privileged groups dominate the oppressed group, the former are inclusive of culture and male and later consist of nature, animal and women.

This chapter is an analysis of Indira Goswami's *The Man from Chinnamasata* (2006) from two ecofeminists stances: the binary apposition between culture and nature (Plumwood 1993) and the concept of animals as absent referents (Adams 2010). This chapter demonstrates the heart-rending agony of animal sacrifice – absent referent – in the name of culture in the Kamakhya temple, which is situated on the top of Nilachal Hill very close to Guwahati, the capital of Assam.

Indira Goswami was a prolific writer, an activist, an editor, a novelist and a professor from Assam. She has been the recipient of prestigious awards like the Jnanpith Award and the Sahitya Akademi Award for her literary works. She has written many novels and short stories, which give glimpses of Assamese life. The notable among them are *The Moth Eaten Howdah of a Tusker* (2005), *Pages Stained with Blood* (2010) and *The Man from Chinnamasta* (2006). *The Moth Eaten Howdah of a Tusker* (2005) pictures the plight of Assamese Brahmin widows. This novel was made into a film, *Adajya*, which won several national and international film awards and was serialised on television too.

The Man from Chinnamasta (2006) was originally written in Asomiya and later translated into English by Prashant Goswami. It is prominently characterised as a propaganda novel against a thousand-year-old ritual of animal sacrifice at the Kamakhya temple. The novel can be studied at three different levels. First, the central character – the great ascetic jatadhari – worships Ma Chinnamasta, and his basic concern is to stop animal sacrifice at the Kamakhya temple even at the cost of his own life. Therefore, he starts a signature campaign – a petition to be signed by the people and presented to the chief priest of the temple for his approval – to stop animal sacrifice with the help of Ratnadhar (the son of the priest Manmohan). On the basis of his in-depth knowledge of Hindu scriptures – the *Puranas* and the *Vedas*, especially, *Kalika Purana* – he gives a true picture of the Hindu ways of the worship of the goddess. The goddess never demands blood either of animals or human beings to appease her thirst. All the way through the novel, Goswami portrays her agitation to stop animal sacrifice through her mouthpiece, that is, the jatadhari, whose fervent appeal is to save animals from man's selfish motive.

Second, on the contrary, the *tantriks*, the *purohits* and the priests of the temple firmly believe in animal sacrifice to worship the goddess. They think this is the only way to worship the goddess. Third, Dorothy Brown, whose husband falls in love with a woman of Kashi tribe and has an illegitimate child, breaks her marital bond with her husband. She becomes a disciple of jatadhari to attain peace. Another important dimension of the novel is Bidhibala, the daughter of Singhdatta Sarma, a friend of Manmohan, who sacrifices her life to save the buffalo.

5.2 The mythological background

The Man from Chinnamasta (2006) portrays a conventional history of the traditions and the rituals of the most famous Kamakhya temple, which depicts the dramatic urgency for change. Before going to analyse the novel, let us probe into the mythological and cultural background of the novel in a nutshell. The very place of Kamakhya has a cultural significance in the history of Assam. Kamakhya is one of the unique temples of the *Shakti Peetha* in India. It is situated on the hillock Nilachal. The mythological legend portrays that the king Daksha organised a *yaga* and for which he invited all the kings, the Gods and the Goddesses except Shiva, his son-in-law. However, Sati, a daughter of Daksha and a wife of Shiva, turned up at the ceremony, where the king Daksha humiliated Shiva. Sati could not tolerate the insult to her husband; she consigned herself to the *yaga* fire. The furious Shiva started the *Tandavanrithya*, the dance of destruction, with the dead body of Sati on his shoulder, which caused havoc among the Gods. The task of soothing Shiva was on the shoulders of Vishnu, who followed Shiva and began to cut the dead body of Sati into pieces with his *Chakra*. These pieces fell on the earth at 51 places, which are recognised as *Shakti Peethas*. The yoni, or female genitalia, fell at the Nilachal or the Kamgiri hill, and this place is established as the Kamakhya temple.

Another legend says that the demon king Narkasura fell in love with the goddess Kamakhya and proposed marriage to her. Kamakhya Devi was puzzled and played a smart trick. She asked him to build a road overnight. Narkasura accepted the condition and he was about to complete the task. But the astute Kamakhya Devi got a cock to announce the daybreak and Naraka's desire was unfulfilled. Moreover, according to the Bodos of Assam the word '*Kham-Maikha*' means 'eater of raw flesh.' In the distant past, the temple was a symbol of tribal fertility rites. In the Middle Ages it stood for *tantric* worship. *Ambubachi* and other ceremonies are still prevalent today. For many centuries, the Kamakhya temple has become an important place for Hindu and Buddhist *tantric* activities. The goddess of the Kamakhya temple is worshipped through animal sacrifice and in the past there were instances of human sacrifice as well. It is substantiated by a continuous discussion of the ancient Sanskrit scriptures like *Yogini Tantra* and *Kalika Purana* in the novel. Both the texts are dedicated to the worship of Kaali and Kamakhya. *Yogini Tantra* is important in the *Vamachara* form of *tantric* worship and contains some historical information, which contributes to explore the novel. *Kalika Purana* is a Hindu religious text, and one of the 18 *Upapuranas*. It describes in detail the rivers and the mountain at *Kamarupatirtha*. It explains in detail the ritual procedures required to worship Kali. By referring to different Sanskrit scriptures, Goswami encapsulates that all religious texts give an indirect message that animal sacrifice is not the way to worship the goddess.

The animal sacrifice is the central narrative thread in Goswami's novel. In the preface to the novel, Dr Prafulla Kotoky refers to it as a successful example of 'Temple Literature' and describes it as 'Kamakhya Lore.' The novel is in the form of vignettes that appear to be disjointed, but it gives variegated kaleidoscopic views of various rituals and activities of the temple.

5.3 The ethnography of animal sacrifice

The Man from Chinnamasta (2006) is not a fictitious one to entertain the readers. However, it has germinated very much from the Indian soil and from Goswami's sheer personal experience. Malaya Khaund in her book *Indira Goswami: A Critical Study of Her Writings* (1994) says, "The greatness of a novel lies mainly on two points, – authenticity and familiarity of the Subject matter. . . . All her novels are products of intimate personal experience and familiarity of the subject matter" (Khaund 1994:1). This novel aims to create awareness against animal sacrifice and it is an empirical rational strategy to change social attitudes. Goswami makes a fervent appeal to stop animal sacrifice in the Kamakhya temple. She has vented out her personal suffocation through this novel; this personal ethos can be seen in her autobiography – *An Unfinished Autobiography* (2002).

Here, it is interesting to note an anecdote, Indira Goswami's mother consulted the astrologers to know Indira's horoscope for marriage. The astrologers predicted that her horoscope was not good. Therefore, bad days lay ahead. A pundit from the Navagraha hills said to her mother, "Better to cut her into two and set her afloat in the river than give her in marriage" (Goswami 2002:16). The pundit's words represent patriarchal views towards woman and her marginalisation; however, the astrologer from the Navagraha hills projects that she had potentiality and certainly could scale the heights. This would be a significant shift – shift from a margin to the centre. To get rid of all evils and the influence of the bad stars, her mother accompanied Indira and her sisters, aunts, the other close relations and well-wishers to sacrifice a goat to the goddess Vagala, who dwells on the top of the Kamakhya hills. As Indira Goswami was sitting in front of the goddess for worship in the form of animal sacrifice, her mind was neither on the goddess, nor on the act of worship. She was very much disturbed and says:

> It bled in agony at the fate of the goat grazing freely over there at the moment. And no sooner had a priest gone out with falchion in hand; my cheeks were flooded with tears. Then all of a sudden, I knew not how it happened; I found myself crying aloud disconsolately in the temple. Everyone was shocked; I heard some outcry of grief and pain. I felt a shivering, warm sensation as someone put a blob of thick blood of the sacrificed goat on my forehead.
>
> (Goswami 2002:18)

This incident always gnaws at and marks a deep scar on her heart. From this incident, she was never healed. This novel is an outlet to bleed the wound.

Second, the reading of the novel *The Man from Chinnamasta* (2006) stirs a feeling in me to study the empirical background to the novel not for just getting an answer to the two research questions but for my concern to nature and animals. To understand the cultural background to the dualistic attitude, I visited Kamakhya temple on 17 February 2013 early in the morning. In the temple, we can take the *Darshan* of the goddess by two ways: one is a special *Darshan*, and the other way is standing in the queue for our turn. I wanted to stand in the queue and take the *Darshan* so that I could closely observe the activities in the temple. I was in the queue for more than three hours. In the meantime, I witnessed continuously for three hours sacrifices of goats, pigeons and hens. There was no alternative for me to avoid watching the pathetic condition of animals, because next to my queue there was the altar of animal sacrifice. The priest was continuously beheading the animals. Even a man was appointed just to sweep the blood. It was a holocaust, which numbed me for a long time. Later, I interviewed the *Pandas* (Priests) and asked them regarding animal sacrifice. They said that there is a ritual that before opening the door of the temple in the morning, they sacrifice an animal to the goddess and then they open the door. Later, the devotees can offer animals to the goddess and this has been followed for ages. Now, we have got an elaborate cultural background to the novel. In the next section, let us examine the elements of ecofeminism.

5.4 Patriarchy and animal sacrifice

A synoptic overview of the Assamese culture is very neatly extrapolated in the novel through different characters. However, no characters are explored deeply; it is like an art gallery or glass menagerie to display artefacts. Let us have a glimpse at the supremacy of the male domination through certain characters who support animal sacrifice: the hermits and *tantriks* of the village Torsa, Singhdatta Sarma, Haladhar Purohit and Shambu Sikadar.

The hermit from the village Torsa comes to the Kamakhya temple once in a year during *Ambubachi* – the annual celebration when the temple is closed for four days because it is believed that the goddess is through her menstrual period – and stays in the Bhairavi temple, which is the premises to butcher the animals. The hermit's inhumane attitude is exposed through the animal's plight: "The animal tried to break free as it was being hauled away to the slaughter house. It wanted to escape the death that came in the form of pilgrims. But the harbingers of doom kept tugging at it. Shoving. Yelling. Prodding" (Goswami 2006:15).

The hermit is a staunch supporter of animal sacrifice. He is against jatadhari's protest march in which more than 200 people joined the rally and signed the memorandum against animal sacrifice. The hermit from Torsa pours out his vengeance and disapproval against all of them,: "The curse of Chinnamasta Devi will blast you to oblivion. Your endeavours will come to naught. The very plan you are hatching to stop animal sacrifice will turn on you like a sword of slaughter. Your blood will flow on the sacrificial altar" (55). Two *tantriks* join their voices with that of the hermit, propelling that man to follow the rituals and scriptures strictly or else they have to meet their doom and bad consequences. "This sorry state of affairs in the country is all because human sacrifice has been prohibited. Evil spirits have taken charge of the land. Only he who is prepared to sacrifice his life to the goddess will be redeemed of his sins. Human blood is the nectar of the goddess. He who offers himself to her is lord to the universe" (56).

With the single stroke of the machete, the *tantrik* beheads the head of the goat and smears the hot blood on his forehead; moreover, he licks the blood. The *tantrik* is very angry about the protest to stop animal sacrifice by jatadhari. Here is what the *tantirk* says:

> This is the first sacrifice in three days. The mother is thirsty. Come, paint your foreheads with this blood. Listen to me. Stab the man from Chinnamasta who tries to rob the Mother of her share of blood. Who will volunteer to stab him! Speak up!. . . . Sacrifice alone will lead you to heaven. Mark my words. A buffalo's blood quenches the goddess's thirst for one hundred years. And when a follower offers the blood of his own body, she is satisfied for one thousand years.
>
> (93)

The *tantrik*, the one who doesn't know the Hindu scriptures, exploits people by misguiding and dominating them:

> You fools! The Goddess Kamakhya has quenched her thirst for thousands of years with blood of both humans and animals. Haven't you heard that

> Ganesha is venerated with liquor, Vishnu with clarified butter, Shiva with music, and Chandika with blood? Are you not familiar with the intricacies of tantric ritual? You dare to change the practice of millennia! You will not. Rot in hell.
>
> (182)

Further, the *tantrik* narrates the story of the king Rudrasingha, who sacrifices thousands of buffaloes for his prosperity. The example of the king Rudrasingha shows that animal sacrifice has passed from generation to generation in the name of culture.

When Haladhar Purohit comes to know that the devotee is not ready to sacrifice the buffalo then he boils down to say, "Scoundrels! You will burn for your impertinence. The sacred texts very clearly state that the blood of a deer satiates the almighty goddess for eight months. The blood of a black bull or a boar appeases her for twelve years" (81). Next, Shambhu Sikdar, whose job is to behead animals for sacrifice, is very proud of his job. He wants to be more proficient so he goes to the riverbank and practices. He picks a grapefruit from the vine and throws into water. As it bobs in water with one stroke of his machete, he slices it into two halves. He slices animals one after the other just like grapefruits. His indifferent inhuman action and a feeling of pride for his profession are shown through his answer to the question of Ratnadhar: "I know! My father said that you have beheaded two thousand buffaloes." Then he answered, "Two thousand and thirty" (89).

Singhdatta Sarma, the father of Bidhibala, comes to Kamakhya to sacrifice a buffalo. He brings a buffalo in the boat. The buffalo was just a baby when his mother was sacrificed at the altar of Ma Kamakhya because Bidhibala's brother was ill. But, unfortunately, Bidhibala's brother did not survive. Now, this buffalo is going to be sacrificed for Bidhibala's marriage. When Singhdatta Sarma comes to know that the buffalo is missing, he becomes very furious and says, "A buffalo has disappeared. I, Singhdatta will bring ten more buffaloes for the ten-armed mother. I shall wash her feet with blood for Bidhibala's sake" (118). The proud patriarchal feeling of Singhdatta Sarma is shown here, "I had promised to sacrifice a buffalo. Now I vow to offer two buffaloes to the goddess. I will sell my land if I have to. I swear" (120). Singhdatta works very hard to sacrifice the buffalo to the goddess. Every day he smears the buffalo's neck with butter because the head would be severed in a single stroke or else it is considered as a bad omen: "a buffalo offered by a Brahmin family from Maravitha had had to be struck three times. The buffalo had been offered to save the life of the man's son who was suffering from cholera. The boy died" (119). He strongly believes in animal sacrifice and wants to sacrifice a buffalo at any cost. Singhdatta loses his temper when he comes to know that Bidhibala had freed the buffalo. He goes out of his way, grabs Bidhibala by her hair and begins to kick her angrily.

Further, the height of man's exploitation and cruelty to animals is very hard to digest. The *deodha* drinks the blood of a dove and dances on the altar of the

Goddess, "Then, without warning, the *deodha* sank his teeth into the goat's neck. As the goat bleated, struggling to free itself, drops of blood fell on the ground. No one there had ever seen anything like this before" (153). In this context, we have to meditate on Carol J. Adams' statement, "The absent referent permits us to forget about the animal as an independent entity; it also enables us to resist efforts to make animals present" (Adams 2010:66). She propounds that the patriarchal values make animals as an absent referents. The patriarchal attitude is revealed through a historical anecdote in the form of a painting: "King Rudrasingha sacrificed twenty thousand buffalos to propitiate the Goddess Durga and flooded this holy abode with blood" (165).

Goswami shows the patriarchal attitude by giving one more example. This is paradoxical. A cord is tied around an elderly man's neck as it is tied around the neck of an animal. The man doesn't utter anything and he collects alms in order to find salvation from his sin of killing a cow, "He had to now atone for his sin by begging for alms with a rope tied around his neck for twelve long years" (24). This superstitious belief is passed from generation to generation. The other example is that the officers of the East Bengal Company – the government of Assam – Captain Welsh and his huge army go to Kalipur, which is the tigers' domain. All these officers go to Kalipur for two reasons: one, for target practice, and two, for hunting, "The sahibs from the Steamer Company cycled up at dawn for target practice. Many a tree had been felled. . . . On occasion a bird dropped from the sky" (28).

The cultural belief of the people is that if they sacrifice buffaloes, goats and ducks on the Goddess's altar, they can get rid of all their woes and worries in their lives. Moreover, Ma Chinnamasta will bless them with a happy life. The people of nearby villages offer animals to the goddess on the occasion of marriages, celebrations and even in their difficult times, "The child played with the plump stippled male goat. Soon the child's forehead would be smeared with his playmate's blood" (32). We can throw light on a few examples that show how man considers animals as absent referents: "Today, countless goats had been brought for sacrifice. The goats were tied up on the veranda of the house in front of the sacrificial altar. The blood from the goats that were sacrificed spattered all over the living goats as well" (138) and "It was shrouded by hundreds of doves smeared with turmeric and sindur. A pile of lifeless goat heads lay at the goddess' feet" (152). Thus, this is a pathetic tale of mute sufferers.

5.5 Women's concern for animals

Another prominent ecofeminist element prevalent in the novel is women as saviours of nature. This is very vividly depicted through the characters such as Bidhibala, Dorothy Brown and an old woman of the village. Bidhibala, the daughter of Singhdatta Sarma, is a mute sufferer. She is going to be married to an old man who has two children. Bidhibala along with her father, brother and grandmother go to Kamakhya to offer a buffalo for her marriage. Bidhibala's short stay in Ratnadhar's house paves the way for her attraction towards

Ratnadhar. Ratnadhar feels pity for her because her marriage is fixed with the old man. Bidhibala likes Ratnadhar and even Ratnadhar has passion for her. Ratnadhar takes her to the Darbhang House to meet Dorothy Brown. He shows paintings of the plight of animal sacrifice at the Kamakhya temple. In the meantime, they come across the painting of the king Rudrasingha's death. There is a tornado in the heart of Ratnadhar. This downpour of feeling is exhibited very minutely in his painting, "What difference was there between sacrificial animals and women? He wanted to warn her. Don't go to the altar Bidhibala. Tell them this wedding can't . . . will not take place" (99). There is a painting that Ratnadhar has not shown to anyone – in which he has poured out his heart and while painting tears rolled down his cheeks – but Bidhibala by chance looks at the painting, and she says, "It wants to live! Oh Ma let it live" (100). This painting penetrates deep into the heart of Bidhibala.

Bidhibala's heart continuously pounds with the thought of the sacrifice of the buffalo for her marriage. The buffalo was brought home, when it was a baby. She has seen the calf growing up before her eyes. She has a lot of attachment to the buffalo. When the thought that the buffalo is going to be sacrificed comes to her mind, her heart shivers:

> Only because of her. Because her marriage had been fixed . . . her heart pounded. She prayed to the goddess to take her life and spare the animal. Its cry was desperate. What must it be thinking? Should she go and check? Suddenly she could see the predatory face of the grey haired man to whom she would soon be married. He had shared the same bed with another woman for ten years.
>
> (102)

She gets up in the mid night and comes near the buffalo. She pats the buffalo and wipes its back with her *achal*. The buffalo trembles, and tears roll down from its eyes. The attachment of the animal with Bidhibala cannot be explicated in the words of the patriarchy. For a moment, she thinks to run away like Dorothy Brown from her people to take shelter with the jatadhari to save the animal from sacrifice. She requests Ratnadhar, who is working day and night to stop the animal sacrifice along with the jatadhari, "Listen carefully. People say you are the one who secretly releases the animals brought for sacrifice. Please let this buffalo go so I won't have to watch it being sacrificed. I haven't slept for nights at the thought. I've seen it grow up, seen its grey tufts turn black. When I bring her food and call 'Mena,' she runs up to me. I can show you now" (104).

For the sake of the buffalo, she is ready to do anything. Therefore, she earnestly requests Ratnadhar, "Ratnadhar, I shall offer my songs to the goddess. Songs created from my tearful, unspoken words. I shall smash the stones weighing down my heart and offer their dust to the goddess along with flowers" (105). When Bidhibala sees her father's smear the animal's neck with butter for one stroke sacrifice, her heart peels out: "She used to wonder if it be better if she died herself. She had seriously considered placing her own head on the sacrificial alter" (120). Being revolutionary, she relieves the buffalo from its doom. She plunges into the darkness and first time dares to raise her head in front of her father and says, "You cannot get buffaloes. You will not bring buffaloes. I shall not marry a man already" (121). After a lot of commotion and fight for the buffalo's loss, Bidhibala leaves the house and meets with a prostitute group (Shekhadari). She ends her life. Bidhibala's concern for animals is extrapolated by the prostitute's cry, "Mother, she would eat neither a morsel nor drink a drop of water the past month. We are prostitutes from Shekhadari – but we did not force her into our trade. She came to us herself. She just cries over and over again. My buffalo has been sacrificed. My buffalo has been sacrificed" (154).

The woman's concern for the animal is very clearly depicted through Bidhibala's life. Bidhibala sacrifices her life to save the animal. Her grim fate after her death is shown through her father's indifference to her death. "Her father, will surely not be willing to perform the last rites for a daughter who has been associated with the North Shekhadari women. Do whatever needs to be done, but do it fast" (156). Ratnadhar's remorse is portrayed through his words, "Bidhibala, they made a sacrifice of you instead of the buffalo" (156).

Another character is Dorothy Brown. She always struggles and strives to stop the animal sacrifice by supporting Ratnadhar to work in her house for a signature campaign. The next character is an old woman. She watches the sacrifice of animals; she throws herself on the sacrificial altar and starts crying and shouting, "A little while ago it was eating grass and leaves. Why did you kill the helpless soul? It was alive. O you blood-thirsty goddess, take my head as well" (93).

5.6 The uniqueness of *The Man from Chinnamasta* as an ecohumanist narrative

The analysis of all selected novels projects the impact of the modern developmental attitude on human beings and nature. However, this novel departs in its thematic concern compared to the other novels selected for analyses. It focuses on animal sacrifice. Further, Goswami enunciates that it is not only women who are struggling to save animals but also men strive hard to save animals. Men also have the same concern for animals and, so, they too are saviours of animals and nature.

In the last section, we discussed how Bidhibala sacrifices her life for the sake of the buffalo. Subsequently, in this section, we examine how men too are saviours of animals. The jatadhari, an ascetic and a devotee of Ma Chinnamasta, is well read of all Hindu scriptures such as the *Vedas*, the *Puranas*, and the *Kalika Purana*. Moreover, he is well-versed in the *Kalika Purana*. On the basis of his in-depth knowledge, he strives to tell the people that Ma has not asked to quench her thirst by the blood of animals: "Child! A child can't be blamed. Ma! Ma! Ma Chinnamasta! Ma! Ma! Read the seventy sixth chapter of the *Kalika Purana* that contains the rules and rituals of worship of the Goddess Shakti and promotes the *Vedas*" (25).

Further, he says that Ma can be worshipped with flowers and other offerings. He is very much against the animal sacrifice. The entire novel is tinged with his protest. He always prays to the goddess imploring, "Ma . . . Ma . . . Ma! Cast off your blood stained robes" (53). His ecofeminist stance is displayed through his ideology and his harmonious understanding about man and nature. "Man is god's creation. Man has many things to learn from animals. Only when men and animals live in harmony will the world become paradise" (180).

In addition, he says, "The mother will no longer wear bloodstained garments. Deck her with flowers. Cover her in garlands of kathal champa, lotus, kanak, kaminikanchan, karabi" (86). The jatadhari's project to stop animal sacrifice influences students and one of the students says, "There will be no more bloodshed on the Mother's premises. Those who want to see an end to animal sacrifice must put their signatures in this book. . . . The scriptures offer alternatives to sacrifice. We can also please the Mother with honey, milk and yogurt. It doesn't say anywhere that the rituals cannot be performed without blood" (87–88).

Ratnadhar, a son of the temple priest Manmohan Sarma, also supports jatadhari in his endeavour to stop animal sacrifice. He always regrets animals' pathetic condition. When he watches a buffalo being dragged to the sacrificial altar, he is attacked by fits and loses his mental balance. In his attack of fits, he throws himself in front of the buffalo, shouting to save it, "Stop! Stop! Don't you see? It's terrified, it doesn't want to go with you" (10). Ratnadhar starts his campaign to stop the animal sacrifice under the guidance of jatadhari. Ratnadhar with the help of more than 200 people starts rallying to spread jatadhari's message to stop the animal sacrifice to the Cotton College students in the valleys. They have planned for the signature campaign. For this, they have prepared a memorandum and wanted to submit it to the head priest. Though Ratnadhar is not well-versed in the scriptures, his inner heart always proclaims to stop animal sacrifice. It is shown, "Yesterday I had a dream, the Mother Goddess whispered in my ear. The fragrance of flowers from her body touched my nostrils. The scent of kharikajara, togor, bakul and lemon flowers" (77). However, his association with the jatadhari has given him some knowledge of scriptures, "Look up the twenty-fifth incantation of the sixty-seventh chapter of the *Kalika Purana*. The white gourd, melon, sugar cane and alcohol are as dear to the goddess as goat's blood" (105).

Ratnadhar vents his feelings and concern for animals through his paintings. "People suspect that he is the one who's been untying the animals brought in for sacrifice" (82). Ratnadhar whole-heartedly works to stop animal sacrifice:

> Human sacrifice stopped only after it was legally suppressed in 1835. How long would it be before the Britishers banned animal sacrifice as well? Who in the world could stand up to their power? Their taxes had sent so many into bankruptcy – but anyone raised a voice in protest? Even Gandhi was said for a ban on animal sacrifice at the Dakshineswar temple.
>
> (82)

He meets almost all the people of nearby villages including the nearby colleges with these words, "You must be aware of our signature campaign. Please put your valuable signature on our book. Once we have the signatures, the white men will help us" (109). Ratnadhar continuously works with the students: "Ratnadhar sat down with a Cotton College student to check that the names and addresses were in line with the thumb impressions. Countless names and addresses. Signatures in Bengali, Hindi and even English. Two students went off to meet a munshi of the court to work out how these signatures would be presented" (116–117). Ratnadhar's remorse and hatred to the animal sacrifice is at its pinnacle when he comes to know that Bidhibala sacrifices her life to save the buffalo. Her sacrifice penetrates into his heart.

In addition, the scripture *Yogini Tantra* describes the ritual to perform Shakti worship in Kamakhya and the other sacred places of Kamarup. It posits "the status of flowers is higher than blood. The sacred texts state that the goddess is satiated for a hundred years with the blood of a single buffalo. The same writings also claim that an offering of one karabi flower can earn the devotee the virtues of the most arduous yagna, the Ashwamedha or horse sacrifice . . . and secure a place in the land of the sun" (129). By spreading the importance of the Holy Scriptures, jatadhari and Ratnadhar strive to stop the animal sacrifice. At the end of the novel, the jatadhari is ready to sacrifice himself in order to save the animals from sacrifice and the ritual of animal sacrifice, as he sliced a piece of flesh from his body:

> He went forward to stand five yards from the chief priest at the sacrificial altar. All the eyes were on him. His massive dreadlocks were the colour of old rusted iron. He raised his powerful arms. Like the mighty Ravan who grew in strength with every wound in battle, he drew a razor from his waistband for all to see, sliced off a piece of his own flesh from below his navel. Holding his bleeding flesh in one hand, he called, "Ma! Ma!"
>
> (186)

5.7 Conclusion

The forgoing analysis shows one more aspect of ecofeminism that is animal sacrifice. In the name of cultural practice, people sacrifice many animals – buffaloes, goats and ducks in the Kamakhya temple, in the belief that they may get the blessings of the goddess. Through the novel, Indira Goswami poses an ethical question to the readers "whether the tradition of shedding animal blood in the name of culture is right." The novel is a signature camp that every reader has to sign on her plea to stop animal sacrifice. Goswami's efforts are aimed to resolve the dualist attitude – culture/nature – and bring animals in the realm of human beings. To quote Dr Kotoky again, "While dealing a sensitive subject like animal sacrifice, the author entirely depends on reason, thoughtful analysis and concurrence of the scriptures. She is not taking an extreme position, nor has she hurt anybody's sentiment" (Afterword). Thus, the novel is an epitome of human beings concern to save the animals which as per our hypothesis, we can consider it as ecohumanism, as Jean-Jacques Rousseau, a Genevan philosopher, articulates, "we may well imagine they ought likewise to partake of the benefit of natural law, and that man owes them a certain kind of duty. In fact, it seems that, if I am obliged not to injure any being like myself, it is not so much because he is a reasonable being, as because he is a sensible being" (Gaard 2011:171).

Bibliography

Adams, Carol. *Neither Man Nor Beast: Feminism and the Defense of Animals*. Continuum, 1994.

———. *The Sexual Politics of Meat*. Continuum, 2010.

Gaard, Greta. *Ecofeminism and Native American Cultures in Ecofeminism*. Temple University Press, 1993b.

———. "Ecofeminism Revisited: Rejecting Essentialism and Re-Placing Species in a Material Feminist Environmentalism." *Feminist Formations*, vol.23, no.2, 2011, pp. 26–53.

———, editor. *Ecofeminism: Women, Animals, Nature*. Temple University Press, 1993a.

Gaard, Greta, and P.D. Murphy, editors. *Ecofeminist Literary Criticism: Theory, Interpretation, Pedagogy*. University of Illinois Press, 1998.

Goswami, Indira. *Pages Stained With Blood*. Betascript Publishers, 2010.

———. *The Man from Chinnamasta*. Translated by Prashant Goswami. Katha, 2006.

———. *The Moth-Eaten Howdah of a Tusker*. Rupa, 2005.

———. *An Unfinished Autobiography*. Sterling Publishers, 2002.

Khaund, Malaya. *Indira Goswami: A Critical Study of Her Writings*. B.R. Publishing, 1994.

Kheel, Marti. "Befriending the Beast." *Creation*, 1987, pp. 11–12.

Plumwood, Val. *Feminism and the Mastery of Nature*. Routledge, 1993.

6 Reconceptualising ecofeminism

From ecofeminism to ecohumanism

6.1 Introduction

The aim of the book, as stated in the introductory chapter, is to examine the ecofeminist discourse in Indian context through the Indian literary narratives. As a result, the study identified several ecofeminist concerns in the novels under examination. Though the Indian novels exhibit the traits of ecofeminism in their thematic concerns and vision, the underlying structure of these novels while portraying the ecological crisis goes beyond the gender issue. After examining the select novels in the previous chapters, the study found that the Indian novels conceptualise the ecological crisis more as a human problem than as a gender problem – woman along with nature are exploited by patriarchy. Therefore, in the present chapter, after giving an overview of the ecofeminist concerns of the select novels, an attempt will be made to reconceptualise the very idea of ecofeminism, which may further be developed by future researchers.

6.2 An overview of the ecofeminists' concern

Ecofeminism has stemmed out of various academic endeavours and activist movements by women against destruction of environment across the globe with the diagnostic focus on women and nature. Ecofeminism is the outcome of ecological crisis induced by scientific revolution, capitalist patriarchy, industrial agriculture and the growth of market culture. It is a discourse, and it is basically oriented to liberate women and nature from the oppression of patriarchal development attitude; it is a quilt theory as it is structurally pluralistic, rather than structurally reductionist or unitary: it emerges from a multiplicity voices, especially women's voices across cross-cultural context (Warren 1994:188). The French feminist Francoise d'Eaubonne coined the term 'ecofeminism' in her ground-breaking work *Le Féminisme ou La Mort* (1974) (*Feminism or Death*) with the core objective to give a call to women to save the planet. To her, "Ecofeminism will put an end to patriarchy and save human society from the devastation wrought on the environment, the nuclear threat and the profit-based system which is at the origin of all war and exploitation on this planet" (D'Eaubonne 1974:4).

The critical interpretation of the major ecofeminists perspectives put forth the elemental reasons for the subjugation of women and exploitation of nature: patriarchy, the basic contention of the exploitation of planet, has paved a way to all destructive attitude in the form of advancement in science and technology, capitalism, market economy, production to consumption, and so on; "technology innovation, the spread of the capitalist market, the scientific revolution, and changing attitudes towards nature and the earth" (Merchant 1980:43) and the development attitude as a reduction paradigm are mal-developments. "Development has meant the ecological and cultural rupture of bonds with nature, and within society, it has meant the transformation of organic communities into groups of uprooted and alienated individuals searching for abstract identities" (Shiva 2010:99). The baseline of this identification is the ideology of dualisms: "These are key ones for western thought, and reflect the major forms of oppression in western culture. Dualisms such as reason/nature may be ancient, but others such as human/nature and subject/object are associated especially with modern, post-enlightenment consciousness" (Plumwood 1993:43) and "Ecofeminism is about deconstructing these dualisms, both in regard to women and in regard to nature" (Ruether 38). Not only the dualisms are the causes for the domination of women and nature but also -isms of domination: "The case is the same for sexism, racism, classism, naturism, and any other 'isms of domination' based on faulty belief systems: what I have called oppressive and patriarchal conceptual frameworks" (Warren 1994:193). Further, the extensive perspective emphasises that it "encompasses a variety of theoretical, practical, and critical efforts to understand and resist the interrelated domination of women and nature" (Eaton and Lorentzen 2003:1).

Even ecofeminists focus on all humankind in all their diversity: "All ecofeminists believe human beings connected to one another and to the nonhuman world: animal, plant, and inert" (Putnam 2009:209) and "needs to start from an understanding of the social relations underpinning current patterns of unsustainability together with an understanding of the material relations between humanity and nature" (Mellor 2000:108). However, in spite of knowing the bondage between human beings and nature, their inclination is towards women and nature.

On the basis of these theoretical underpinnings, it, very precisely, connotes that the emergence of ecofeminism is from the ecological and feminists movements, the core objective of ecofeminism is to preserve nature by reconstructing the existing social structure, "of one's interests in ecological protection, preservation, and reconstruction; and of one's 'special' ecological consciousness" (Sandilands 1999:5) and also "environmental ethics can benefit by incorporating feminist insights on the limitations of traditional philosophical conceptions of ethics (Cuomo 1998:i). It highlights a few alternative projects for preservation, for example, though history has established a connection between women and nature by considering women as a lower to culture and men, and all social practices is the root of ecological destruction, but this nature/culture dualism can be resolved by dialectical feminism (King 1983, 1995), ecological feminism (Warren 1994), feminist environmentalism (Agarwal 1998) and so on. It also "creates support in favour of alternative development models (as opposed to market-oriented capitalist ones)" (Datar 2011:2).

The cursory glance of all these ecofeminists put forth three basic tenets such as one, patriarchal development attitude is the fundamental reason for the subjugation of women and exploitation of nature (Ruether 1975; Daly 1978; Griffin 1980; Mies & Shiva 1993; Mies 1986); two, basically there is a connection between women and nature based on the basis of oppression of both. Though some ecofeminists reject the connection between women and nature (Ortner 1974; King 1983) and the other says the connection between women and nature is strength (Daly 1978), the most of ecofeminists opine that there is a connection between women and nature on the basis of ideology, biology (Starhawk 2002), ontology (Griffin 1980) and the oppression of the both (D'Eaubonne 1974; Ruether 1975; Agarwal 1998; Merchant 1980; Adams 1994) therefore, women are closely associated with nature; three, women are life giver, nurturer, caretaker, sustainer of nature, and responsive to environmental issues (D'Eaubonne 1974; Ruether 1975; Daly 1978; Merchant 1980; Mies & Shiva 1993). These aspects are theoretical underpinning for the analysis of the novel.

6.2.1 What are the major thematic aspects of ecofeminism articulated in the Indian novels?

The contextualisation of the Indian novels within the framework of ecofeminism delineates these three basic aspects. We can trace the first core aspect of ecofeminism – the modern developmental attitude – in the Indian novelists' narration. Kamala Markandaya's *Nectar in a Sieve* (2009) focuses on the introduction of tannery in a rural setting which impacts on the agrarian culture. On the parallel lines, Pundalik Naik's *The Upheaval* (2002) represents the emergence of mining in Kolamba village due to which all the villagers are forced to give up their occupation of cultivating their ancestral lands and join the mining work. The mining work not only grabs the occupation of villagers but also pollutes the land and contaminates the rivers. In a similar vein, Shivram Karanth's *Return to Earth* (2002) portrays the effects of colonial modernisation on the agrarian culture. Sarah Joseph's *Gift in Green* (2011) exhibits the impact of construction of a megacity on agriculture and water bodies.

Kamala Markandaya's *The Coffer Dams* (2008) projects the impact of dam construction on the habitat of animals and tribal people; Na D'Souza's *Dweepa* (2013) describes the agony of the displaced local people due to the problem of submersion of the land for the construction of the dam. Indra Sinha's *Animal's People* (2007) tells about the worst effect of the accidentally released deadly toxic gas from the pesticide factory of Bhopal on nature and human beings.

Indira Goswami's *The Man from Chinnamasta* (2006) is a heart-rending agony of animal sacrifice in the name of culture. This novel depicts two ecofeminists stances: a binary apposition with culture and nature, and the concept of animals as absent referents.

The second key aspect of ecofeminism is 'women as saviours and nurturers of nature.' This concept can be traced through Rukmani, the protagonist of the novel *Nectar in a Sieve* (2009). She represents the Third World women peasants and their attitude towards sustainable agriculture with their traditional agrarian knowledge of cultivation of the land. In the same vein, the concept is registered through a few women characters such as Kunjimathu, Shailaja, Karthiayani, Devaki, Divya, Gitanjali and others, who fight to save the water bodies and paddy fields of Aathi in *Gift in Green* (2011). Further, Helen, the central character of the novel *The Coffer Dam* (2008), projects feminine qualities of caring and nurturing. She always has an intimate association and bonding with the tribal people and nature. Then, Karanth, in *Return to Earth* (2002), postulates women's conservationist quality through their various domestic and day-to-day activities. The women in the novel spend most of their time either looking after the fields or taking care of the cattle. Paraothi and Sarasothi look after the cattle by preparing gruel by mixing hay and bran, milking the cow, driving out the cattle to the fields, and after that gathering the cow-dung from the cowshed to throw into the manure pit. Next, Naik (2002) projects a woman's concern for nature through Rukmini's protest against her husband's desire to join the mining for earning. Rukmini is very much associated with her field. Initially, she is not ready to accept the mining in place of their ancestral occupation of farming.

The third key aspect of ecofeminism propose that the causes of ecological destruction and exploitation of women are capitalist patriarchy and scientific revolution. For example, due to the industrial disaster, in *Animal's People* (2007), the women characters like Aliya, Payré Bai, Lilabai, Gargi and others suffer a lot. Further, in *The Upheaval* (2002) the victims of the mining are Rukmini and Kesar; in *Nectar in a Sieve* (2009) the tannery affects the life of Rukmani, whereas the sufferers of colonial modernisation are Paraothi, Sarasothi and Nagaveni in *Return to Earth* (2002).

Along with these key aspects of ecofeminism, the Indian novelists have attempted to delineate a few more aspects; for example, the impact of consumer culture on the pastoral culture. As depicted in *Nectar in a Sieve* (2009), the commercial culture grabs the occupation of agriculture. The next aspect of ecofeminism is that most novels deal with is the concern for the agricultural life. Further, the analysis shows that Indian peasants and tribal people deeply revere their occupation either related to the land or forest as they exhibit divine faith in these. In *The Upheaval* (2002), the reverence for the land and river is shown through the celebration of agriculture-oriented festivals and worship of the river. These are a few aspects of ecofeminism that Indian novelists have tried to narrate in their novels.

6.2.2 Do Indian novels reproduce the concept of ecofeminism as developed in the West or do they treat the problem in a different way?

In the last section, we dealt with the main elements of ecofeminism in the Indian novels. This section throws light on how the select Indian novels provide new insights into the ecofeminist discourse; do they reproduce the concept of ecofeminism, or do they treat the problem in a different way? In the previous chapters, we examined the uniqueness of each novel in their treatment of the ecological crisis. As most novels testify, they look at the crisis more as a human problem than merely as a gender problem. In this respect, they show that men too are saviours of nature and equally subjugated by environmental disasters.

Several examples may be cited from these novels. The tribal old headman in the novel *The Coffer Dams* (2008) is preoccupied with the impact of dam construction on the habitat of animals and tribal people, and he is not able to resist the involvement of tribal people in the construction of dam. In the novel *The Upheaval*, Abu, a 60-year-old man of the village Kolamba, is disappointed with the villagers' leaving behind their age-old occupation of cultivating the land and joining the mining. In every possible way, Abu wants to save the land from pollution and the Mandovi River from contamination. He says, "Would our people forsake their fields to go and break stones at the mines otherwise?" (Naik 68). Yet, another very important character like Abu is Salvo Master, who has an ardent desire to teach the children the impact of mining on human beings and nature. In every possible way, he wants to save nature. The basic intention of *Return to Earth's* is to critique the fad of the villagers to send their children to English schools (the colonial modernisation). However, at the end, the novel depicts the return journey of Rama from the city to the village. His desire to work on the land plays a vital role in the novel. The novel portrays his concern for land by leaving behind the glamorous world of city where he works.

Similarly, Sarah Joseph's *Gift in Green* (2011) demonstrates Dinakaran's concern for animals and birds. Dinakaran has taken action on the bird trafficking. He gives his life for the sake of the waterbodies and paddy fields of his village. Along with Dinakaran, there are a few more male characters continuously struggling to save the paddy fields and water bodies like Noor Mohammed, Chandramohan, Ponmani and Markose. Subsequently, in *The Man from Chinnamasta* (2006), the jatadhari, an ascetic and a devotee of Ma Chinnamasta, strives to stop the animal sacrifice in the name of culture. The jatadhari propounds that all religious texts give an indirect message that animal sacrifice is not the way of worshipping the goddess. Ratnadhar a son of the temple priest Manmohan Sarma, also supports the jatadhari in his endeavour to stop animal sacrifice. He starts a campaign to stop animal sacrifice under the guidance of the jatadhari, which is called the signature campaign. Further, Nathan, from Kamala Markandaya's *Nectar in a Sieve* (2009), wishes that his sons work on the land and follow the agrarian culture for which he has a divine passion and attachment. He does not accept the introduction of tannery in place of his agrarian culture. Likewise, in Na D'Souza's *Dweepa* (2013) Ganapayya is against the developmental attitude as he says that if the trees are cut down for railway tracks, highways, telegraph lines, dams and townships, global warming would be exacerbated and there would be less rain.

Further, these novels show that men are also subjugated by ecological crisis. In Kamala Markandaya's novel, despite spending their entire life working very hard on the land, Rukmani and Nathan are left with nothing in the end. Being forced to join a stone quarry in the city, Nathan meets his tragic death. Nathan is thus lost his life because the introduction of tannery in the village. In *The Upheaval* (2002), the continuous carrying of loads of chips and pollution caused by mining leads to Pandhari contracting tuberculosis. Nanu meets his death at the mining due to the slide of a huge mound of mineral dust. Similarly, in *Animal's People* (2007), the first-hand victims of the industrial disaster include Aliya's grandfather, Hanif; Shambhu, who can hardly breathe; Yusuf Omar, who develops an ulcer; and Rafi, who is seriously ill. In *The Coffer Dams* (2008), the premature blast during the dam construction has taken a toll of 28 tribal men. Thus, these are a few special and unique qualities of the Indian novels, which add a new dimension to the ecofeminist discourse

6.3 Reconceptualising ecofeminism: from feminism to humanism

The select Indian novels reflect the essence of Indian environmental problems. They are germinated from the soil of India and tinged with the novelists' personal experiential ethos. The dominant mode of the narration in these novels is realism. The task of the novelists is not only to document the facts and figures but also to recreate the real human experience artistically. A close examination of the Indian novels based on the writers' empirical understanding of environmental crisis reveals that it is appropriate to look at the environmental problems from human angle without getting into gender issues. Although these novels exhibit feministic concerns, they are, at the same time, not blind to men as saviours of nature and subjugated by ecological crisis. Thereby, we can infer that feminism, which has largely been imported from the West, takes a backseat and ecological crisis appears as a general human problem in Indian novels. They represent ecological problems from the point of view of a general human problem, not just as a feminine problem. Further, these Indian novelists' narration of environmental problems of India compels us to rethink these theoretical formulations of ecofeminism which segregate woman and nature from men. That is, the environmental problems are not merely feminine problems but they are human problems. Therefore, this book proposes to reconsider ecofeminism as ecohumanism.

There is a lot of debate on humanism too. It has been a contested arena. Structuralism and post-structuralism have questioned the validity of the notion of humanism. This book does not intend to get into such a debate. To understand humanism, we need to explore the conceptual understanding of humanism. The Western philosophers and thinkers – Protagoras, Socrates, Plato and Aristotle – proclaimed that man is supreme. The historical evolution of humanism has shown subtle development in the understanding of its philosophy. The philosophers and thinkers of the Renaissance period focused on the problems related to the dignity of man and immortality of truth. Later, it emphasised reason, passion, values and potentialities of human beings that lead to an intellectual revolution. This revolution gave a way to develop various forms of humanism, for example, religion of humanity, democratic humanism and scientific humanism (Syamala 1998). It is very apt to have an intent look at Stephan Law's opinion: "Humanism means a system of thought in which human values, interests, and dignity are considered especially important" (Law 1).

On the other hand, the genesis of Indian humanism can be traced back to the Indian Holy Scriptures such as the *Vedas* and the *Bhagavad Gita*. These are concerned with human life through their uplifting from the physical level to self-conscious spiritual level. Further, Jainism says that man has got the power to work out his ways without the assistance of the supernatural. Buddhism believes that man has the ability to realise human values. The modern Indian humanism is influenced by the Western humanism, which says that man is supreme. The modern Indian philosophers and thinkers – Swami Vivekananda, Rabindranath Tagore and Mahatma Gandhi – concentrate on the problems of foreign domination, economic development and social evils along with the harmonic relation between man and man, man and nature, man and universal spirit and his religion (Syamala 1998).

Though there is a yawning gap between the Western and Indian philosophical understanding of humanism, the basic orientation of humanism is to solve human problems. However, there is a shift in the comprehension of humanism because of the advancement in science and technology and civilisation, which transcended its limited conceptual understanding. The shift gives a wider dimension to the existing understanding of humanism. It propounds that there is a world beyond anthropocentrism, which is inclusive of all natural species, which is called post-humanism. The basic objective of post-humanism is to claim that human beings have no right to destroy nature. Moreover, they are part of nature, not above nature. The discourse of post-humanism rejects anthropocentric dominance. Further, it advocates human life is one among many lives. We, at present, have reached the post-human thinking, where man is no more the centre of our discourse. Placed in this context, a wide range of scholarship and research has contributed to this discourse: Halberstam's and Livingston's *Posthuman Bodies* (1995), Francis Fukuyama's *Our Posthuman Future: Consequences of the Biotechnological Revolution* (2002), MacCormack's *Posthuman Ethics* (2012) and Rosi Braidotti's *The Posthuman* (2013). The basic contention of these polemics is that advancement of science and technology and economic globalisation are the vital causes for the serious disruption of ecosystem.

All these posthuman theorists sum up that if the advancement accelerates in this pace, the future of all species is in danger and therefore needs a sustainable relationship of human and non-human that is decentralising human beings human central attitude. However, it does not mean that it is the end of humanism. Although there are different interpretations of the concept of humanism (for example, 16th- and 17th-century Europe witnessed a renewal of interest in scholarship and it is known as the Renaissance Humanism), the present study focuses on the centrality of human beings in different discourses. Ecohumanism is defined as against the politics of gender identity. It stresses the centrality of human beings, not just the gendered human being. Therefore, considering the Indian novelists' insights, we can suggest an alternative approach that is ecohumanism.

Further, if we look back at the ecofeminist discourse, we will find diverse critical opinions on it. As Plumwood states, "An examination of the literature shows many serious difficulties in the position as so far stated, and that some forms of ecofeminism are untenable or open to serious objection. These are the major gaps in the arguments of the position, a need to clarify many of the key concepts and to distinguish more carefully between quite different positions which have been lumped together under the ecofeminism label" (Plumwood 120). Here, it is very apt to quote the remarks of the editor of *Signs* (a leading journal of academic feminism) on the rejection of the publication of a review essay on ecofeminism in June 1992, "This is really an opinion piece [and] the ties to women are not very clear." In the same year, in the month of November, the editors of the academically reputed journal *NWSA* too had rejected the essay as well as, "[i]t is a difficult argument to make – what does concern with ecology have to do with concern with sexism, racism, and classism?" (Gaard,2011:33). It is true that the oppression of women and nature is because of the development of science and technology, modernisation, globalisation, industrialisation and so on; however, these modern developmental attitudes are not only exploiting women and nature but human beings as a whole.

6.4 Conceptual framework

Why should we care for ecofeminism? We have increasingly been addressing the environmental problems across the globe. It needs no elaboration as to how international and national agencies including nation-states and NGOs are struggling to deal with the contemporary environmental crisis. Accordingly, one of the notable features of our time is the rise and proliferation of recognising the status of women in our life. It is critically important to take care of both environment and women. Hence, our engagement with ecofeminism makes sense either in the form of activism or in the form of intellectual practice. This book is an intellectual engagement with the idea of ecofeminism. Activism has its own ways of dealing with things, changing this world. But true changes require powerful ideas to make it an active agent of change. So, thinking is also a practice that participates in the politics of change. Ecofeminism as an intellectual movement in the human sciences makes an attempt to understand what is happening to us and how to make use of this understanding to lead a good life.

In this respect, the Western theorisation of ecofeminism cannot be accepted as a universal understanding of the ecological crisis. The present work, while recognising the problem of environment in India, questions the hegemony of the Western culture in thinking about the human problems. Placed in this context, this chapter tried to answer two questions. The answer of the first question focuses on the treatment of ecofeminism in Indian novels. These novels exhibit the major concerns of the standard ecofeminist discourse such as the critique of patriarchal development attitude, women as saviours and subjugated by environmental disaster, etc. However, this gives us a partial picture of the idea of the environment problem. The answer to the second question, in what way Indian novels add a new dimension to the ecofeminist discourse, shows that it is appropriate to look at the environmental crises more as a general human problem than merely as a gender problem. We can argue that it is appropriate to look at the issue from human angle, without getting into gender issues. Although these novels exhibit feministic concerns, they are equally not blind to men as saviours and subjugated by ecological crisis. Hence, after the critical examination of the select Indian novels, the present work tried to point out that India looks at the environment problem differently.

Therefore, one of the conclusions of the present study is that Indian narratives on ecological crisis look at the environmental problem more as a human problem than the problem of feminism. If the ecofeminism is the dominant trope in the Western discourse, ecohumanism is the dominant discourse in India. It does not mean that the Indian narratives are not concerned with the problem of women. They do show the elements of yoking the problems of nature with the problems of women, but they rarely involve either in the polemics or in the identity politics of feminist writings bring to the environment problem. Hence, the present study proposes to reconceptualise ecofeminism as ecohumanism. Thus, the research defines ecohumanism as, "an equivalence relation of man and woman by dissolving dualistic approach to resolve ecological crisis through common consensus." Ecohumanism propounds that human beings – both women and men – and nature are part of a complex web. This research does not claim to have theorised this concept fully, but it has identified an important trait of ecological problems in India.

Another interesting finding of the present study is that the Indian environment discourse, especially the novels that are examined in the study, offer the critique of the Western development attitude. Instead of focusing on the patriarchy, they hold the Western development responsible for the environmental crisis. According to these narratives, the West is the source of the markers of development and exploitation.

Finally, the book makes a modest contribution to the ecofeminist discourse in general and to the Indian novel in particular. Although it is an ecofeminist reading of the Indian novels, the study is not a simple exercise in applying the concepts of ecofeminist theory to the Indian novels. It is an attempt to engage with the theory of ecofeminism in the light of reading of the Indian novels. Application of theories, as noted by Sheldon Pollock, does not advance knowledge. Instead, we need to test the theories (Pollock 2006). In this respect, the present study made a modest attempt at testing the ecofeminist theory that has come from the West. In scrutinising the theory, the study came to a conclusion that the notion of ecofeminism needs reconceptualization, and ecohumanism is a better theoretical understanding that works for the Indian culture. Obviously, the Indian novel has not been studied from the ecofeminist point of view, and the present study fills that gap.

Bibliography

Abrams, M.H. *A Handbook of Literary Terms*. Cengage Learning India, 2009.

Adams, Carol. *Neither Man nor Beast: Feminism and the Defense of Animals*. Continuum, 1994.

———. *The Sexual Politics of Meat*. Continuum, 2010.

Agarwal, Bina. "Environmental Management, Equity and Ecofeminism: Debating India's Experience." *Journal of Peasant Studies*, vol.25, no.4, 1998, pp. 55–95.

Bowler, Peter J. *The Fontana History of the Environmental Sciences*. Fontana Press, 1992.

Braidotti, Rosi. *The Posthuman*. Polity Press, 2013.

Cuomo, Chris. *Feminism and Ecological Communities*. Routledge, 1998.

Daly, Mary. *Gyn/Ecology: The Metaethics of Radical Feminism*. Beacon Press, 1978.

Datar, Chhaya. *Ecofeminism Revisited*. Rawat Publications, 2011.

D'Eaubonne, Francoise. *Le feminism ou la mort*. Pierre Horay, 1974.

———. *Philosophical Writings*. Translated and edited by Norman Kemp Smith. Macmillan, 1952.

———. "What Could an Ecofeminist Society Be?" *Ethics and the Environment*, vol.4, no.2, 1999, pp. 179–184.

Derrida, Jacques. *Structure, Sign and Play in the Discourse of the Human Sciences*. Translated by Alan Bass. Routledge, 2003.

D'Souza, Na. *Dweepa*. Translated by Susheela Punitha. Oxford University Press, 2013.

Eaton, Heather, and Lois Ann Lorentzen, editors. *Ecofeminism and Globalisation*. Rowman & Littlefield Press, 2003.

Fukuyama, Francis. *Our Posthuman Future: Consequences of the Biotechnological Revolution*. Profile Books, 2002.

Gaard, Greta. *Ecofeminism and Native American Cultures in Ecofeminism*. Temple University Press, 1993b.

———. "Ecofeminism Revisited: Rejecting Essentialism and Re-Placing Species in a Material Feminist Environmentalism." *Feminist Formations*, vol.23, no.2, 2011, pp. 26–53.

———, editor. *Ecofeminism: Women, Animals, Nature*. Temple University Press, 1993a.

Gaard, Greta, and P.D. Murphy, editors. *Ecofeminist Literary Criticism: Theory, Interpretation, Pedagogy*. University of Illinois Press, 1998.

Goswami, Indira. *The Man from Chinnamasta*. Translated by Prashant Goswami. Katha, 2006.

Griffin, Susan. *Women and Nature: The Roaring Inside Her*. Harper & Row Publishers, 1980.

Halberstam, Jack, and Ira Livingston, editors. *Posthuman Bodies*. Indiana University Press, 1995.

Heidegger, Martin. *Being and Time*. Translated by Joan Stambaugh. State University of New York Press, 1956.

Joseph, Sarah. *Gift in Green*. Translated by Valson Thampu. Harper Perennial, 2011.

Karanth, Shivram. *Return to Earth*. Translated by Padma Ramachandra Sharma. Sahitya Akademi Centre for Translation, 2002.

King, Ynestra. "The Eco-Feminist Perspective." *Reclaiming the Earth: Women Speak Out for Life on Earth*, edited by L. Caldecott and S. Leland. Women's Press, 1983.

———. "Engendering a Peaceful Planet: Ecology, Economy, and Ecofeminism in Contemporary Context." *Women's Studies Quarterly: Rethinking Women's Peace Studies*, vol.23, no.3/4, 1995, pp. 15–21.

Law, Stephen. *Humanism: A Very Short Introduction*. Oxford University Press, 2011.

MacCormack, P. *Posthuman Ethics*. Ashgate, 2012.

Markandaya, Kamala. *The Coffer Dams*. Penguin Books, 2008.

———. *Nectar in a Sieve*. Penguin Books, 2009.

Mellor, Mary. "Feminism and Environmental Ethics: A Materialist Perspective." *Ethics and the Environment*, vol.5, no.1, 2000, pp. 107–123.

Merchant, Carolyn. *The Death of Nature*. HarperCollins Publishers, 1980.

Mies, Maria. *Patriarchy and Accumulation on a World Scale*. Zed, 1986.

Mies, Maria, and Vandana Shiva. *Ecofeminism*. Rawat Publications, 1993.

Naik, Pundalik. *The Upheaval*. Translated by Vidya Pai. Oxford University Press, 2002.

Ortner, Sherry B. "Is Female to Male as Nature Is to Culture?" *Women, Culture, and Society*, edited by M.Z. Rosaldo and L. Lamphere. Stanford University Press, 1974.

Plumwood, Val. "Ecofeminism: An Overview and Discussion of Positions and Arguments." *Australian Journal of Philosophy*, vol.64, 1986, pp. 120–138.

———. *Feminism and the Mastery of Nature*. Routledge, 1993.

Pollock, Sheldon. *The Language of the Gods in the World of Men*. Permanent Black, 2006.

Ruether, Rosemary Radford. "Ecofeminism: First and Third World Women." *American Journal of Theology & Philosophy*, vol.18, no.1, 1997, pp. 33–45.

———. *New Women/New Earth: Sexist Ideologies and Human Liberation*. Seabury, 1975.

Salleh, Ariel. "Class, Race and Gender Discourse in Ecofeminism and Deep Ecology Debate." *Environmental Ethics*, vol.15, no.3, 1993, pp. 225–244.

Sandilands, Catriona. *The Good Natured Feminist*. University of Minnesota, 1999.

Shiva, Vandana. *Biopiracy*. Natraj Publishers, 2012.

———. *Staying Alive Women, Ecology and Survival in India*. Women Unlimited, 2010.

———. *The Violence of the Green Revolution*. Zed Books Publication, 1993.

Sinha, Indra. *Animal's People*. Simon & Schuster, 2007.

Starhawk. *Webs of Power: Notes from the Global Uprising*. New Society Publishers, 2002.

Syamala, K. "Humanisms in Modern Thought with Special Reference to the Philosophy of Jawaharlal Nehru." Diss. University of Calicut, 1998.

Tong, Rosemarie Putnam. *Feminist Thought: A More Comprehensive Introduction*. Westview Press, 2009.

Warren, J. Karen, editor. *Ecofeminist Philosophy: A Western Perspective on What It Is and Why It Matters*. Rowman & Littlefield, 2000.

———. "Ecofeminist Philosophy and Deep Ecology." *Philosophical Dialogues*, edited by Nina Witoszek and Andrew Brennan. Rowman & Littlefield, 1999.

———, editor. *Ecofeminism: Women, Culture, Nature*. Indiana University Press, 1997.

———. *Ecological Feminism*. Routledge, 1994.

———, editor. *Ecological Feminist Philosophies*. Indiana University Press, 1996.

———. "The Power and Promise of Ecological Feminism." *Environmental Ethics*, vol.12, no.3, 1990, pp. 125–146.

Weeks, Marcus. *Philosophy in Minutes*. Quercus Editions, 2014.

Wolfe, Cary. *What Is Posthumanism?* University of Minnesota Press, 2010.

Index

absent referent 126, 132, 141
agrarian culture 8, 21, 38, 39, 41, 47, 48, 50–53, 56, 57, 59, 62–64, 69, 70, 84, 87, 140, 142, 143
agrarian profession 40
androcentric attitude 20, 92, 101
animal sacrifice 7, 8, 126, 127–131, 133–136, 141, 142
anthropocentric attitude 5, 91, 101
Aranya Kaanda 27
Aranyanisukta 27

capitalist patriarchy 49, 93, 138, 141
colonial modernisation 37, 62, 63, 67, 70, 87, 140–142
conceptual claim 16
conservation of resources 17
corporate agriculture 15, 32, 78, 87

deep ecology 11, 12
deforestation 12, 15, 30, 31
development model 22, 140
displacement 3, 19, 31, 90, 103, 104, 110

ecocriticism 8, 12, 13
ecohumanism 6, 87, 137, 138, 143, 145, 146
ecohumanist exposition 85
ecohumanist narrative 47, 58, 69, 110, 134
ecological concern 2, 27, 98
ecological crisis 4, 6, 7, 8, 13, 19, 26, 32, 37, 82, 88, 89, 90, 91, 112, 138, 142, 143, 146
ecological destruction 14, 141
ecological problems 6, 7, 16, 26, 59, 96, 143, 146
ecological sensibility 100
ecological studies 2
ecopsychology 12
ecosophy 11
empirical claim 15
environmental deterioration 15
environmental justice 8, 12
environmental movements 29, 30, 32, 72
environmental narratives 28
environmental studies 1, 2
epic tradition 26

gender analysis 30
globalisation 7, 14, 20, 23, 32, 65, 112, 144

Hindu discourse 22
homogeneous system 21
human problem 6, 59, 69, 110, 123, 138, 142, 143, 144, 146
humanism 6, 62, 143–145

ideological shifts 21
impact of dam construction 3, 7, 19, 89, 91, 105, 140, 142
impact of tannery 7, 37, 38, 45, 46, 47
Indian agriculture 17, 48, 56, 61, 70
Indian context 1, 6, 7, 11, 21, 26, 32, 61, 138
Indian scriptures 26
indigenous people 5, 20, 21
industrial disaster 8, 112–115, 118, 120–125, 141, 143
industrialisation 7, 11, 13, 57, 92, 123, 145

landfilling 8, 11, 72, 81, 84, 86
literary narratives 6, 8, 11, 26, 28, 29

mining industry 50
mixed cropping 21, 43, 68, 106

modernity 3, 5, 19, 24, 37, 62
monoculture 4, 21, 31
mythological background 8, 127

narratives of agriculture 37
nature-centric life 12
new positional ecology 12

Oshadhi Sukta 27

pastoral life 37, 89
patriarchal developmental attitude 63, 65, 73, 74, 77, 95, 97, 105, 113
patriarchal power 2, 13, 32
patriarchy 2, 32, 49, 53, 93, 99, 129, 133, 138, 139, 141
peasants and tribes 101
plight of native community 104
praxis-oriented 2, 32

realism 6, 37, 143
reductionist agriculture 23
regional ecology 5
regional novels 4, 26
renewable agriculture 17

scientific revolution 11, 21, 40, 101, 138, 139, 141
shallow ecology 12
social concerns 12, 24
socio-economic movements 14
soil fertility 17, 21, 56, 106
sustainable agriculture 19, 141

technology and science 3, 19
tribal community 3, 19
tribal people 3, 5, 22, 46, 62, 89, 90–99, 100, 101, 102, 110, 140–142

uncivilised people 101

Vanaparvan 27
Vanavaasa 27

women-led protest 14

For Product Safety Concerns and Information please contact our EU representative GPSR@taylorandfrancis.com
Taylor & Francis Verlag GmbH, Kaufingerstraße 24, 80331 München, Germany

www.ingramcontent.com/pod-product-compliance
Lightning Source LLC
LaVergne TN
LVHW021037011026
849160LV00009BA/679
* 9 7 8 1 0 3 2 0 9 1 1 8 1 *